John Herman Merivale

Notes and formulae for mining students

John Herman Merivale

Notes and formulae for mining students

ISBN/EAN: 9783337214722

Printed in Europe, USA, Canada, Australia, Japan

Cover: Foto ©Paul-Georg Meister /pixelio.de

More available books at **www.hansebooks.com**

NOTES AND FORMULÆ
FOR
MINING STUDENTS.

PREFACE.

The following pages do not profess to contain much original matter. They are a collection of notes and formulæ, drawn from various sources, my authority being quoted in most instances, and were originally compiled for the students in the Durham College of Science, because I could find no suitable text-book at a moderate cost. They are now re-issued, revised, and enlarged, in a form which I trust may be useful not only to students but to my professional brethren.

The principal sources of information upon mining matters are the Transactions of the various engineering societies to which the student, in most of our large towns, has access. I have given a great many references to the most familiar of them, so that the student, who wishes to follow up a subject, may be in a position to acquaint himself with details which want of space does not permit me to include in a work like this.

The examples of the use of the formulæ, which I have added at the end of the book, are merely given to assist students working without a teacher, and are not intended to furnish practical designs for mining appliances.

I should like to take this opportunity of thanking my colleagues for the valuable assistance they have given me in the revision of my Notes.

J. H. M.

NOTE TO THE SECOND EDITION.

I HAVE taken the opportunity of the demand for a New Edition to correct a few verbal inaccuracies, and to substitute for the paragraphs relating to the Mines Act of 1872 others relating to the present Act.

J. H. M.

CONTENTS.

COURSE OF STUDY FOR MINING STUDENTS.

Mathematics.—Experimental physics.—Chemistry.—Geology.—Mechanical drawing.—Modern languages.—Certificated Managers' Examination. 1

COAL FIELDS OF GREAT BRITAIN AND IRELAND.

Formation of coal.—Table of sedimentary rocks.—Mining statistics 6

NEWCASTLE COAL FIELD.

Section of the strata.—List of seams.—Dykes.—Troubles.—Mineral products.—Drawings and persons employed.—Coalfields of the world.—Composition of coal.—Produce of coal seams.—Weight of ores.—The Coal Commission . . . 10

THE STRENGTH OF MATERIALS.

Ropes.—Chains.—Cast-iron pipes.—Boilers.—Masonry pillars.—Beams.—Girders.—Columns.—Dams.—Tubbing . . . 16

TIMBER.

Timber measure.—Cost of timbering 26

EXPLOSIVES.

The Mines Act and general rules.—The Explosives Act.—Composition of explosives.—Substitutes for explosives . . . 28

MACHINERY.

Nature and uses.—Force, work, power, and energy.—Table of specific heats **32**

PRIME MOVERS.

Men and horses.—Cost of feeding horses.—Work of men and horses.—Hydraulic motors.—Windmills.—The steam-engine.—Horse-power **35**

BOILERS.

Strength.—Boiler fittings.—Explosions.—Incrustations.—Doctors.—Corrosions.—Horse-power of boilers.—Properties of steam.—Condensation.—Chimneys **45**

TRANSMISSION OF POWER.

Compressed air.—Wire ropes.—Steam.—Loss by radiation and loss by contact with air in horizontal pipes and in vertical pipes.—Loss by friction.—Steam transmission.—Practical details.—Summary **54**

MAGNETISM AND ELECTRICITY.

Practical electro-magnetic units.—Compass surveying.—Firing shots.—Signalling.—Lighting.—Transmission of power.—Firedamp detectors.—Danger **67**

SEARCH FOR MINERALS.

Costeaning.—Boring.—Hand-boring.—The diamond process.—Cost of boring.—Deep bore-holes **71**

SINKING.

Mines Act.—Workmen in the pit, Ordinary method and Poetsch method.—Workmen at bank, Kind-Chaudron method and Chavette method.—Shaft fittings.—Shaft pillars.—Deep mines.—Underground temperature.—Important Adits . . **74**

SYSTEM OF WORKING.

Above-ground.—Under-ground.—Long wall and bord and pillar.—Stoping.—Cost of working 79

WINDING.

Winding engines.—Counterbalances.—Pulley frames.—Pulleys.—Ropes.—Chains and cages.—Sundries 85

DRAINING.

Pumps.—The shaft.—The workings.—Sundries.—Boring against old workings.—Dams.—Tubbing. 89

HAULING.

Resistances to be overcome, Friction, Inclination, Curves.—Motors, men, horses.—Self-acting inclines.—Cost . . 95

GENERAL PROPERTIES OF AIR AND GASES.

Pressure.—Temperature.—Occluded Gases.—Transpiration.—Diffusion.—Fire-damp analyses 101

CHEMISTRY.

Compounds and elements.—Atoms.—Chemical symbols.—Molecules and formulæ.—Chemical equations 106

THE GASES.

Oxygen.—Carbonic oxide.—Hydrogen.—Hydrogen-sulphide.—Nitrogen.—Carbonic acid.—Fire-damp.—Ambulance.—Asphyxia.—Illuminating gas.—Air 109

VENTILATION.

The three laws of friction.—Causes of difference of pressure.—Depressive ventilation.—The furnace.—Consumption of fuel in furnaces.—The steam jet.—Centrifugal fans.—Compresive ventilation.—Relation between volume, pressure, &c.—Table of hyperbolic logarithms, and table of the specific gravities of solids. 122

EXAMPLES OF THE USE OF FORMULÆ.

PAGE

Ropes.—Strength of boilers.—Size of dams.—Resistance to traction.—Self-acting incline.—Position of meetings.—Pumping engine.—Boilers required for engine and consumption of coal.—Dimensions of chimney.—Units of work to compress air.—To find dip and angle of bed by three boreholes.—Ventilating pressure required to circulate air.—Ventilating pressure produced by furnace.—Power of hauling engine.—Effect of cupola.—Size of pipe required to transmit steam. . 136

INDEX 153

ABBREVIATIONS USED THROUGHOUT THE BOOK.

Trans. N.E.I. = Transactions of the N. of England Inst. of Mining and Mechanical Engineers.

Proc. I.M.E. = Proceedings of the Inst. of Mechanical Engineers.

Proc. I.C.E. = Proceedings of the Inst. of Civil Engineers.

RANKINE M.M. = RANKINE's Machinery and Mill Work.

RANKINE S.E. = RANKINE's Steam Engine.

E.M.F. = Electro-motive force.

HP = Horse-power.

NOTES, FORMULÆ, &c.,

FOR

MINING STUDENTS.

INTRODUCTION.

COURSE OF STUDY FOR MINING STUDENTS.

I AM frequently asked by students to advise them as to the course of study they should pursue in order to pass the Colliery Managers' examination; or, more generally, to enable them to become properly qualified mining engineers.

So far as the certificate is concerned, I have always found (premising, of course, a thorough knowledge of reading, writing, and arithmetic), that an intimate acquaintance with all the details of a colliery, both above and below ground, is of much more importance than book learning. The student should, however, have the Mines Act, General and Special Rules at his fingers' ends, and such a knowledge of the theory of the gases and ventilation as may be obtained from Atkinson's book. But, of the man who aspires to a high place amongst the mining engineers of the next generation, something more than this is required. Every year the necessity of a thorough grounding in certain departments of science is becoming more apparent, and I have no hesitation in saying that the successful English mining engineer of the future will have to be (as his continental rival long has been), not only a good practical man, but a man of science as well.

I would advise, then, a good general education up to the age of about sixteen, such as would enable him to pass the examination in general knowledge now required by most of the professions. He should then go through the following course of study which can be pursued in one or other of the science colleges :—

Mathematics.

Euclid.—Books I., II., III., IV., VI., XI., and the definitions of Book V.

Arithmetic.

Algebra as far as the Binomial Theorem.

Trigonometry, including the solution of triangles and the use of logarithms.

The laws and principles of *Mechanics*, and the properties of *Conic Sections* treated Geometrically.

Analytical Geometry of two dimensions.

The elements of the *Differential and Integral Calculus*, with one independent variable.

Statics and Dynamics treating of forces in one plane.

The elementary parts of *Hydrostatics* treated mathematically.

Experimental Physics.

Elementary Mechanics, Heat, Sound, Light, and Electricity.

Advanced mechanics, Heat, Magnetism, and Electricity; special attention being paid to the strength of materials, the construction and use of Hydraulic and Pneumatic Machines, the Measurement of Gases, the Conservation of Energy, the Theory of the Steam Engine, Electric Lighting, Electric Signalling, and the Transmission of Power.

Laboratory Practice.

Chemistry.

General Principles of Chemistry; History of the Non-Metallic Elements.

History of the Metals and their more important native and artificial compounds; Principles of Qualitative Analysis.

Laboratory Practice.

Geology.

Elementary Mineralogy and Lithology (with demonstrations).

Physical Geology.

Stratigraphical Geology and Elementary Palæontology, with Special Reference to the British Isles.

Mining Geology, or Descriptive Geology, with special reference to the Coal-producing and Metalliferous regions of the world.

Geological Surveying.—As far as possible the instruction should be given in the field, and should be of such a nature as to give students a practical knowledge of Elementary Geological Surveying and " prospecting."

Mechanical Drawing.

Plane Geometry.—Construction and use of scales; construction of various polygons; delineation of the various curves; miscellaneous problems relating to lines, circles and plane figures.

Solid Geometry.—Miscellaneous problems of lines and planes represented by plan and elevation, &c. Determination from various data of the projection of solids. Intersections, Isometrical projection.

Constructional Drawing.—Detailed drawings of various examples of building construction.

Mechanical Drawing.—Detailed drawings of working parts. Drawings of Miscellaneous Machines and Engineering Structures. Working and finished drawings. Drawings from models and actual machines.

Modern Languages.

Although a colloquial knowledge of modern languages may not be necessary, the student should at least be able to read French and German mining literature.

Such a course as this would take from two to three years, according to the previous education of the student. It should be begun at about the age of sixteen, prior to the four or five years' practical training.

THE CERTIFICATED MANAGERS' EXAMINATION.

Speaking generally, the candidate must be 22 years of age or upwards, and have had five years' experience at a colliery. The subjects for Examination are Reading, Writing, Arithmetic, the Geology of the district in which the Examination is held, the Mines' and Explosives' Acts and Mine Engineering; but, as each district has its own rules too which are changed from time to time, the candidate should apply to the Secretary.

Mining District.	Chief Inspector. Address and Date of Appointment.	Assistant Inspector. Address and Date of Appointment.	Secretary to Board of Examinations. Address and Date on which Examinations usually held.
Manchester and Ireland.	Joseph Dickinson, South Bank, Pendleton, Manchester. 1850.	J. Turton, jun., 75, Manchester Road, Bolton. 1885.	Maskell W. Peace, King Street, Wigan. December.
North Staffordshire.	Thomas Wynne, Gnosall, Stafford. 1852.	A. R. Sawyer, Basford, Stoke-on-Trent. 1879.	Joseph Knight, Newcastle-under-Lyne. June.
Scotland (Western Division).	J. M. Ronaldson, Pollokshields, Glasgow. 1875. In charge 1886.	G. F. Bell, Hill Head, Glasgow. 1886.	Stuart Foulis, 135, St. Vincent Street, Glasgow. November.
Midland.	A. H. Stokes, Greenhill, Derby, 1874. In charge 1887.	W. H. Hepplewhite, Leicester. 1885.	Wm. Saunders, The Wardwick, Derby. October.
Scotland (Eastern Division).	Ralph Moore, 13, Clairmont Gardens, Glasgow. 1862.	R. McLaren, Coatbridge. 1885.	Robert Calder, 296, Renfrew Street, Glasgow. May.

INSPECTORS, &c., ADDRESSES.

District	Inspector	Assistants
South Wales.	J. T. Robson, 1873. In charge 1887.	F. W. Randall, Penarth, Cardiff. 1883. F. A. Gray, Swansea. 1886. C. H. James, 8, Courtland Terrace, Merthyr Tydfil. January.
Yorkshire.	Frank N. Wardell, Wath-upon-Dearne, Rotherham. 1867.	John Gerard, Wakefield. 1874. James Mellors, Leeds. 1885. John R. Jeffery, 5, Piccadilly, Bradford. June.
Newcastle-upon-Tyne.	James Willis, Newcastle-upon-Tyne. 1871.	J. B. Atkinson, Stocksfield, Newcastle-upon-Tyne. 1873. R. P. W. Oswald, Brigham, Carlisle. 1885. Fred. Gosman, The Mining Institute, Newcastle-upon-Tyne. January.
Durham.	Thomas Bell, Durham. 1873.	W. N. Atkinson, Shincliffe Hall, Durham. 1873. J. Plummer, Bishop Auckland. 1885. G. W. Bartlett, Tees Grange, Darlington. July.
Liverpool.	Henry Hall, Rainhill, Prescot. 1873. In charge 1875.	J. L. Hedley, Chester. 1874. Maskell W. Peace, King Street, Wigan. June.
South Western.	J. S. Martin, Swansea. 1873. In charge 1886.	R. Donald Bain, Newport, Monmouthshire. 1877. J. T. Thomas, Coleford, Gloucester. July.
South Staffordshire.	W. B. Scott, Parkdale, Wolverhampton. 1873. In charge 1883.	W. H. Pickering, Finchfield, Wolverhampton. 1883. W. Blakemore, Shelfield, Walsall. January.

THE COAL FIELDS OF GREAT BRITAIN AND IRELAND.

THE rocks forming the earth's crust are divided by geologists into two groups, viz. :—
The Aqueous or Stratified ;
The Igneous or Unstratified.

These groups are subdivided into numerous systems, and in one of the systems of the stratified group coal is found in such large quantities as to have conferred upon it the title of the Carboniferous System.*

The table following shows the systems of the stratified group and their subdivisions in descending order, beginning with the superficial deposits down to the lowest known depths.

At the commencement of the Lower Carboniferous period (upper old red sandstone) the greater part of the British Isles, south of Perth, appears to have been occupied by two seas, separated the one from the other by a ridge of land passing through the centre of Wales, Shropshire, Worcester-

---

* It must not, however, be supposed that coal is found only in the Carboniferous system. As a matter of fact, it occurs more or less in every one of the Aqueous divisions. For example :—

In the Tertiary we have the lignites worked for many years at Bovey Tracey, in Devonshire ; and the lignites, brown coals, and pitch coals of the continent.

The Cretaceous contains lignite and bituminous coal in Spain, Germany, New Zealand, and South America.

The Jurassic : The Kim-coal of Kimmeridge, the moorland coal of Yorkshire, and the coal of Brora in Sutherlandshire. Besides these, there are many lignites and bituminous coals worked on the continent.

The coal-fields of India, and a part of the Australian coal-fields, are in the Triasso-Jurassic formation.

The Permo-Carboniferous furnishes coal in North America, Bohemia, and Autun in France. In the Devonian are situated some of the coal-fields of N. W. France, as in Mayenne. From the Silurian, coal has been worked in Portugal.

TABLE I.
TABLE OF THE CHIEF DIVISIONS RECOGNIZED IN THE SEDIMENTARY ROCKS OF BRITAIN.

Those Formations which occur in Northumberland and Durham are denoted by asterisks.

TABLE OF STRATIFIED ROCKS.
[N.B.—The figures give *maximum* thicknesses only.]

	PERIODS.	SYSTEMS.	FORMATIONS.	LIFE-PERIODS.	
	QUATERNARY.	POST-TERTIARY or PLEISTOCENE (250 ft.)	*Peat, Cave and Valley-Gravel Deposits. *Brick-earths and Loess. *Raised Beaches, &c. *Boulder-Clay and Gravels.		Dominant type, Man.
Represented locally by a break in the succession.	TERTIARY or CAINOZOIC.	PLIOCENE (100 ft.)	Norwich, Red, and Coralline Crags.	Range of Fossil Birds in time.	Dominant types, Birds and Mammals.
		MIOCENE (125 ft.)	Bovey Beds (?).		
		EOCENE (2,600 ft.)	Fluvio-Marine Series. Bagshot Beds. London Tertiaries.	Range of Fossil Mammals in time.	
	SECONDARY or MESOZOIC.	CRETACEOUS (7,000 ft.)	Maestricht Beds. Chalk. Upper Greensand. Gault. Neocomian. Wealden.	Range of Fossil Reptilia in time.	Dominant type, Reptilia.
		JURASSIC (3,000 ft.)	Purbeck Beds. Portland Beds. Kimmeridge Clay. Coral Rag. Oxford Clay. Great Oolite. Inferior Oolite. Lias.	Range of Fossil Fishes in time. Range of Invertebrata and Plants in time. Footprints of Birds?	
		TRIASSIC (3,000 ft.)	Rhætic. Keuper. Muschelkalk. *Bunter (?).		
Break (unconformity).		PERMIAN (500 to 3,000 ft.)	*Red Sandstone and Marl. *Magnesian Limestone, &c. *Yellow Sand.		Dominant type, Fishes.
Break represented by the Cheviot Rocks.	PRIMARY or PALÆOZOIC.	CARBONIFEROUS (20,000 ft.)	*Coal Measures and Millstone Grit. *Carboniferous Limestone or Bernician Series. *Tuedian and Basement Beds or Up. Old Red Sandstone.		
		DEVONIAN (5,000 to 10,000 ft.)	Devonian. Old Red Sandstone.		Dominant type, INVERTEBRATA.
		*SILURIAN (3,000 to 20,000 ft.)	Upper Silurian. Lower Silurian or Ordovician.		
		CAMBRIAN (20,000 to 30,000 ft.)	Cambrian.		
		PRE-CAMBRIAN, &c. (30,000 ft.)	Pre-Cambrian.		

[G. A. LEBOUR.]

shire, and south Staffordshire, into the Eastern Counties. These seas were gradually filled up with limy ooze, sand, and mud; so that, at the commencement of the Upper Carboniferous period (coal measures), they had become two large swamps. In these swamps the vegetation grew which now forms our seams of coal. Each seam marking a period when the swamps were above water, the intervening beds of sandstone and shale a period when they were below.

At the close of the Carboniferous period the earth's crust in these districts was upheaved along several parallel east and west lines. Taking the North of England as an example, two of these axes of upheaval passed, the one to the north, the other to the south, of the present Newcastle coalfield. The denuding agencies of rain, frost, &c., planed off the tops of the ridges, sweeping away the rocks lying high up in the series, together with the seams of coal; the result being that the Newcastle coalfield occupied the eastern portion of a trough of coal measures extending from the west of Ireland to the German Ocean; and this trough was separated from similar troughs to the north and south (now occupied in part by the coalfields of the south of Scotland and Yorkshire, &c.) by tracts of country denuded of their coal. Upon these troughs of coal measures and barren intervening country the Permian, the next formation in ascending order, was deposited. Again, there was a period of upheaval; but now along lines running north and south. One of these ran to the west of Newcastle, and formed the Pennine chain, or great central ridge of the North of England. Again, the agents of denudation set to work, planed down the arch and separated the Newcastle coalfield from that of Whitehaven.

It is to the intersection of these two series of axes of upheaval, approximately at right angles the one to the other, that the basin shape of our coalfields is due; while the disseverance of these basins the one from the other has been the necessary consequence of the planing down and sweeping away of the arches by the action of rain, ice, &c.

The fields of coal now left to us by denudation are shown in the following table. The names of the districts are not those of the separate basins, but the groups into which they have been divided under the Coal Mines Act.

TABLE II.
MINES INCLUDED UNDER THE COAL MINES' ACT.

NAMES OF DISTRICTS.	PERSONS EMPLOYED.	COAL.	FIRE CLAY.	IRONSTONE.	SHALE, &c.	TOTAL.	MINES.	LIVES LOST.
Northumberland, Cumberland, and North Durham	53,011	16,427,529	320,939	773	5,086	16,754,327	203	91
South Durham, Westmoreland and North Yorkshire	55,771	20,402,479	221,013	—	—	20,623,492	194	78
Cleveland (Ironstone)	6,713	—	—	5,932,244	—	5,932,244	36	10
North and East Lancashire	31,640	9,738,674	96,696	—	90	9,835,460	287	229
Ireland	860	109,035	900	—	—	109,935	27	1
West Lancashire and North Wales	41,565	13,176,944	59,047	2,594	2,098	13,240,683	225	97
Yorkshire	63,562	18,497,778	200,722	126,596	57,753	18,882,849	395	84
Lincolnshire (Ironstone)	115	—	—	81,521	—	81,521	10	1
Derby, Leicester, Nottingham, and Warwickshires	55,163	16,963,684	77,122	29,386	4,051	17,074,243	355	73
North Stafford, Cheshire, and Shropshire	21,889	6,324,600	66,000	1,920,920	15,392	8,326,911	225	46
South Stafford and Worcestershire	23,137	9,862,497	185,632	117,726	2,457	10,168,312	365	29
Monmouth, Somerset, parts of Glamorgan and Breconshire	35,659	9,352,072	88,277	26,037	11,632	9,478,018	294	70
South Wales	61,617	17,207,540	106,920	35,656	6,704	17,356,820	305	224
Scotland, East	46,022	14,905,620	192,826	504,023	1,665,667	17,268,136	323	75
Scotland, West	23,908	6,382,966	262,420	1,331,136	114,487	8,091,009	244	42
Totals in 1885	520,632	159,351,418	1,878,514	10,108,612	1,885,416	173,223,960	3,488	1,150
Do. in 1884	520,376	160,757,779	2,053,927	10,412,443	1,648,610	174,872,759	3,554	942
Do. in 1883	514,933	163,737,327	2,189,452	11,495,401	1,341,210	178,763,390	3,707	1,054

THE NEWCASTLE COALFIELD (NORTHUMBERLAND AND DURHAM).

Section of the Strata.

The Post Tertiary, all four divisions of which are found; but more especially the *Boulder clay*, which consists, for the most part, of stiff blue and brown clays, with boulders of limestone, sandstone, basalt and porphyrite. It covers almost the whole of the coalfield.

The Permian Formation, about 600 feet thick. It extends over the East and South-East of Durham. It rests unconformably upon the coal-measures, and consists of the magnesian limestone, and the yellow sand (often very wet and loose), and produces limestones; but no coal.

The Coal Formation probably extends (either at the surface or beneath more recent formations) over the whole of Durham or Northumberland, excepting the Cheviot district. It may be subdivided into—1. The Upper Coal Measures, from the base of the Permian to the roof of the High Main Seam, about 1,100 feet. Here are found the Hebburn Fell, Five Quarter, and Three Quarter Seams. 2. The Middle Coal Measures, from the roof of the High Main to the floor of the Brockwell, about 900 feet. Here are found all the best * seams of the district, some sixteen in number, of a thickness, in the aggregate, of about 50 feet, discarding those less than 18 inches. Fifty feet, however, will not be found in any one section; but 30 feet may be taken as an average. 3. The Lower Coal Measures and millstone grit formation, from the floor of the Brockwell to the roof of the Fell-top limestone, perhaps 600 feet, contain no seams of any present

---

* Amongst these the following may be mentioned:—The High Main of Northumberland (a house coal) is the Three Quarter of the Hetton district (an inferior coal) and the Shield-row of Pontop (a gas coal).

The Metal coal and the Stone coal of the Tyne run together to form the Grey Seam of Northumberland (a steam coal) and the Five

value. 4. The Bernician, from the roof of the Fell-top limestone to the base of the Harbottle grits, 2,500 to 8,000 feet. Several seams of coal are found here; but they are variable in quality, thickness, and extent. The best known are, perhaps the little limestone coal, and the Shillbottle seam. 5. The Tuedian beds from the base of the Harbottle grits to the Silurian formation.

By the term, "The Newcastle Coal Field," only the upper and middle coal measures are usually meant. These extend approximately from the Coquet to the Tees, and may be divided into—

Exposed coal field	460 square miles.
Beneath the Permian	225 ,,
Beneath the sea	111 ,,
	796 square miles.

It was estimated, in 1870, that the upper and middle coal measures contained	7,452,250,000 tons.
The limestone coal	580,000,000 ,,
	8,032,250,000 ,,

Quarter of the Hetton and Pontop districts (a house coal). The Metal coal produces house coal.

The yard seam of Northumberland (steam) is the Main coal of Hetton (house) and the Brass Thill of Pontop (a gas coal).

The Bensham of Northumberland (steam) is the Maudlin of Hetton (gas and house).

The Five Quarter of Northumberland (steam) is the Low Main of Hetton (steam and house).

The Bensham and Five Quarter run together to form the Hutton Seam of Pontop.

The Low Main of Northumberland (steam) is the Hutton Seam of Hetton (house) and the Main coal of Pontop (coke).

The Beaumont of Northumberland (unworked to the east) is the Harvey of Hetton (gas and coking) and the Towneley of Blaydon (house coal).

The Stone coal and Five Quarter, found in the S.W. of Northumberland (steam and house coal) unite to form the Busty Bank of Pontop (coking and gas coal).

The Brockwell, not worked in E. Northumberland, is a house coal at Blaydon and a coking coal about Brancepeth.

including all seams above 18 inches thick, and not more than 4,000 feet deep. Since that date about 400,000,000 tons have been worked.

Dykes.

The principal whin dykes run east and west, and "are remarkably uniform in lithological composition. They are, as a rule, close-grained basalts, deep blue when freshly broken, and weathering to brown or red." The following may be mentioned beginning from the North :—*Acklington, Bedlington ; Hartley ;* and *Coley-Hill,* in Northumberland ; *Hebburn,* the southern extension of Coley-Hill ; *Hett ;* and *Cockfield* in Durham.

Troubles.

These, too, run for the most part east and west. The most important, beginning from the north, are :—Dipper South, of about 50 fathoms, between Newbiggin and North Seaton ; the *Ninety-fathom Dyke,* a dipper North, which runs from Cullercoats, between Killingworth and Gosforth Collieries south of Newburn, through Whittonstall to a little west of Minsteracres, where it dies out ; *Stublick,* a dipper North, which, starting from a little to the south of Corbridge, runs west into Cumberland ; and *Butterknowle,* from Wingate Grange to Butterknowle, a dipper South of 40 fathoms.

Mineral Products, 1881.

Coal, 35,592,420 tons from 365 collieries. Iron ore, 70,771 tons from Weardale. Fireclay, 438,251 tons, Durham principally. Lead ore, 17,467 tons; Silver, 54,036 oz. ; Pyrites, 5,466 tons; Barytes, 5,435 tons: from the Bernician. Limestone for furnace linings from Permian ; for a flux and agricultural purposes from Permian and Bernician. Basalt for road metal from dykes. Sandstone for building, grindstones, and filters. Cement stones from the Tuedian. Since 1881 Salt has been pumped near Middlesborough ; but there has been a decrease in the production of iron ore, and lead.

The important commercial position of this district is due

—1st. To the numerous seams of coal, which are thick, produce the best coal of every kind, and are cheaply worked. 2nd. To the position of the coal along a coast line, indented with many natural harbours. 3rd. To the near neighbourhood of the Cleveland ironstone (Lias). 4th. To sundry minor advantages, such as the lead, fireclay, and building stones ; and now the salt (Trias, or upper Permian) found in the neighbourhood of Middlesborough, at a depth of about 200 fathoms.

The Drawings and persons employed in Northumberland and Durham during the seven years, 1879—85, have been :—

TABLE III.

		Tons.	Persons Employed.
1879.	Northumberland	5,537,071	22,153
	North Durham	6,105,259	17,900
	South Durham	17,146,644	49,556
1880.	Northumberland	6,850,162	23,048
	North Durham	7,075,846	18,528
	South Durham	20,987,500	53,224
1881.	Northumberland	7,074,577	22,740
	North Durham	6,986,930	18,481
	South Durham	21,530,913	54,810
1882.	Northumberland	7,060,783	23,368
	North Durham	7,458,006	19,529
	South Durham	21,780,808	55,969
1883.	Northumberland	7,527,065	23,793
	North Durham	7,738,870	19,621
	South Durham	22,139,565	57,067
1884.	Northumberland	7,516,005	25,423
	North Durham	7,618,254	20,403
	South Durham	20,934,049	56,533
1885.	Northumberland	7,354,776	26,519
	North Durham	7,340,007	19,712
	South Durham	20,397,317	55,729

TABLE IV.

The Principal Coal Fields of the World.

	Estimated area in square miles.	Estimated thickness in feet.	Produce in 1880, in millions of tons.	Consumption in 1880, in millions of tons.	Produce per man employed in 1880, in tons.
United States	196,000	20	70·3	70·1	
Australia	30,000		1·8		
China	20,000				
British North America (1884)	7,500	20	1·87		
Great Britain	5,500	35	149·3	130·1	337
British India (1883)...	2,500	50	1·31		
Russia	2,000		2·9		
Prussia	1,500		59·2	56·8	295
France	1,000	60	19·4	28·5	166
Belgium	550	60	16·9	12·1	164
Austro-Hungary			16·0	14·6	
New Zealand (1884)			0·48		
New S. Wales (1884)			2·75		

Composition of Coal.

Coal varies in composition as it passes from lignite, the first stage in its production from vegetable matter, through its subsequent stages into anthracite. This is shown in the following table, after Thomas, "Coal, Mine-gases, and Ventilation," p. 6.

TABLE V.

	Carbon.	Hydrogen.	Oxygen.	Nitrogen.	Sulphur.	Ash.
Wood	50·0	6·0	41·5	1·0		1·5
Lignite	56·0	5·0	25·0	1·5	2·0	10·5
House Coal ...	81·0	5·0	8·0	1·5	1·0	3·5
Steam Coal ...	89·0	4·0	2·5	1·0	1·0	2·5
Anthracite ...	91·5	3·0	1·5	1·0	1·0	2·0
Cannel	80·0	6·0	7·0	1·5	1·0	4·5

Produce of Coal Seams, &c.

The produce depends upon the specific gravity of the coal (1·25 to 1·50), the system of working, and the number of faults, balks, etc. A rough and ready rule is to calculate the produce at 100 tons per inch per acre, which leaves an ample allowance (about 25 per cent.) for loss of every kind. The weight in the seam per inch per acre = specific gravity × 101.

At a colliery in Durham, working the Harvey Seam, 3 feet 6 inches thick, 5,185 tons per acre were got when working long wall; 5,052 tons bord and pillar.

TABLE VI.

WEIGHT PER SQUARE FATHOM ONE INCH THICK OF VARIOUS ORES IN POUNDS.

Gold Native	3,281·25	Zinc Red Oxide	1,012·50
Silver do.	1,875·00	Blende Zinc Sulphide	750·00
Copper do.	1,668·75		
Vitreous Copper	1,050·00	Nickel Glance	1,406·25
Copper Pyrites	787·50	Cobalt Glance	937·50
Red Copper	1,106·25	Iron Pyrites	912·75
Malachite (Carbonate of Copper)	712·50	Magnetite	1,016·25
		Arsenical Iron Ore	1,068·75
Tin Oxide (Cassiterite)	1,256·25	Specular do.	912·75
Galena (Lead Sulphide)	1,406·25	Hæmatite	750·00
		Pitch Blende (Uranium)	1,312·50
Lead Carbonate (Cerussite)	1,200·00	Baryta	750·00

The Coal Commission of 1871, estimated that 146,480,000,000 tons of available coal were left in Great Britain and Ireland; including no seams less than one foot thick, or at a greater depth than 4,000 feet. Since then, about 2,000,000,000 tons have been worked.

Authorities:—"Extent and Duration of the Northern Coal Field;" T. Y. Hall, Trans. N.E.I., ii. 104. "Rivers, Ports and Harbours of the Northern Coal Field;" T. Y. Hall, Trans. N.E.I., x. 41. "Outlines of the Geology of Northumberland and Durham," Lebour; "The Coal Fields of Great Britain," Hull. "Coal: Its History and Uses," Green, Miall, etc. Several Papers in the "Trans. N.E.I. Mining Records," Hunt. "The Coal Seams of the Northumberland and Durham Coal Field," J. B. Simpson. The Report of the Coal Commission, 1871. The Reports of the Inspectors of Mines.

THE STRENGTH OF MATERIALS;
Some approximate formulæ.

	STRAIN.	FRACTURE.	
Elementary.	Extension.	Tearing.	(Ropes.)
	Compression.	Crushing.	(Short Columns.)
Compound.	Distortion.	Shearing.	(Rivets.)
	Twisting.	Wrenching.	(Shafts.)
	Bending.	Breaking across.	(Beams.)

The *factor of safety* is the ratio that the breaking strain should bear to the working load; it depends upon the nature of the load and the material, as follows (with some exceptions):—

Material.	Dead Load.	Live Load.
Metal	3	6
Masonry and Brickwork	4	8
Wood and Hemp	5	10

The *proof strain* should be from $\frac{1}{3}$ to $\frac{1}{2}$ the breaking strain; or twice the working load.

Round Ropes.

W = Breaking load in tons.
C = Circumference of rope in inches.

(1.) $W = 0.25 C^2 \therefore C = \sqrt{\dfrac{W}{.25}}$ for hemp ropes.

(2.) $W = 1.50 C^2 \therefore C = \sqrt{\dfrac{W}{1.5}}$ for iron wire ropes.

(3.) $W = 3 C^2 \therefore C = \sqrt{\dfrac{W}{3}}$ for crucible steel wire ropes.

(4.) $W = 4 C^2 \therefore C = \sqrt{\dfrac{W}{4}}$ for improved plough steel wire ropes.

As the sizes of ropes are usually referred to their circumferences, the following formulæ will be useful :—

c = Circumference in inches.
A = Area in square inches.
d = Diameter in inches.

(5.) $c = 3\cdot1416 d = \sqrt{12\cdot5664 A}.$

(6.) $A = \cdot7854\, d^2 = \dfrac{c\,d}{4} = \cdot07958 c^2.$

(7.) $d = \dfrac{c}{3\cdot1416} = \cdot3183 c = \sqrt{\dfrac{A}{\cdot7854}}.$

In calculating the size of rope required to support a given weight, the weight of the rope itself must be taken into account; but the weight of the rope cannot be calculated until its size is known.

If c = circumference of rope in inches.
w = weight roughly in lbs. per fathom.

(8.) $w = \dfrac{c^2}{4}$ hemp.

(9.) $w = \dfrac{c^2}{1\cdot2}$ iron or steel.

Combining these formulæ with formulæ (1), (2), (3), and (4), we get—

(10.) $c = \sqrt{\dfrac{L}{\dfrac{0\cdot25}{M} - \dfrac{F}{4 \times 2240}}}$ for hemp.

(11.) $c = \sqrt{\dfrac{L}{\dfrac{1\cdot5}{M} - \dfrac{F}{1\cdot2 \times 2240}}}$ for iron.

(12.) $c = \sqrt{\dfrac{L}{\dfrac{3}{M} - \dfrac{F}{1\cdot2 \times 2240}}}$ for crucible steel

(13.) $c = \sqrt{\dfrac{L}{\dfrac{4}{M} - \dfrac{F}{1\cdot 2 \times 2240}}}$ for improved plough steel.

Where c = circumference of rope in inches.
L = Load, viz., full cage and chains in tons.
M = Factor of safety (from 6 to 10).
F = Depth of pit in fathoms.

At a certain depth, the weight of the rope will be equal to the safe working load. Thus for round crucible steel ropes with say 8 as factor of safety, we see from formulæ (3) and (12), that the limit of depth in fathoms is:—

$$\dfrac{2240 \dfrac{3c^2}{8}}{\dfrac{c^2}{1\cdot 2}} = 1{,}008 \text{ fathoms.}$$

When the time arrives to work mines at such depths, taper ropes must be used, see formulæ (14) and (15), or the mineral raised in more than one lift.

Round Taper Ropes.

Are made of a decreasing size from the top to the bottom, so that they may be as strong at the top, where the strain is greatest, as they are at the bottom, where the strain is least.

A = Area of rope at any point in square inches.
a = Area of rope at bottom end in square inches.
w = Weight of one cubic inch of the rope in lbs. (For an iron or steel rope w = 0·14, for a hemp rope w = ·043. Both these numbers are approximate only, as the weight depends partly upon the size of the rope.)
L = Safe load in lbs. per square inch of section of rope (say, Iron, 7,000; Steel, 11,500; Plough Steel, 13,440; Hemp, 740).
D = Distance in inches from A to a.
W = Weight of rope in lbs.
e = 2·7182.

(14.) $A = ae^{\frac{wD}{L}}$

(15.) $W = La\left(e^{\frac{wD}{L}} - 1\right) = L(A - a)$.

To avoid the use of logarithms, the following table of the values of $e^{\frac{wD}{L}}$ for distances from 10 to 600 fathoms is given. By it the dimensions of a round plough steel taper rope at points 10 fathoms apart from the bottom to the top can be calculated.

TABLE VII.

Distance from A to a in fms., i.e., $\frac{D}{6 \times 12}$	$e^{\frac{wD}{L}}$	Distance from A to a in fms., i.e., $\frac{D}{6 \times 12}$	$e^{\frac{wD}{L}}$	Distance from A to a in fms., i.e., $\frac{D}{6 \times 12}$	$e^{\frac{wD}{L}}$	Distance from A to a in fms., i.e., $\frac{D}{6 \times 12}$	$e^{\frac{wD}{L}}$
10	1·0075	160	1·1275	310	1·2617	460	1·4119
20	1·0151	170	1·1359	320	1·2712	470	1·4226
30	1·0227	180	1·1445	330	1·2808	480	1·4333
40	1·0304	190	1·1531	340	1·2904	490	1·4441
50	1·0382	200	1·1618	350	1·3001	500	1·4549
60	1·0460	210	1·1705	360	1·3099	510	1·4726
70	1·0539	220	1·1793	370	1·3198	520	1·4769
80	1·0618	230	1·1882	380	1·3284	530	1·4880
90	1·0698	240	1·1972	390	1·3397	540	1·4992
100	1·0778	250	1·2062	400	1·3498	550	1·5105
110	1·0859	260	1·2153	410	1·3600	560	1·5219
120	1·0941	270	1·2244	420	1·3702	570	1·5333
130	1·1024	280	1·2336	430	1·3805	580	1·5449
140	1·1106	290	1·2429	440	1·3909	590	1·5565
150	1·1190	300	1·2523	450	1·4014	600	1·5682

Flat Ropes.

Are formed of two or more round ropes stitched together, and their strength may be calculated accordingly, a deduction being made of about 10 per cent.

Another rule. C. M. Percy, "Mechanical Engineering of Collieries," p. 71, says:—

Round Ropes.
(16.) $C^2 \times 4$ charcoal iron.
(17.) $C^2 \times 6$ crucible steel.
(18.) $C^2 \times 10$ plough steel.

Flat Ropes.
(19.) Width $\times$ thickness $\times 35$ charcoal iron.
(20.) Do. $\times$ do. $\times 55$ crucible steel.
(21.) Do. $\times$ do. $\times 70$ plough steel, gives the safe working load in cwts.; the dimensions are taken in inches.

Chains.
W = Breaking load in tons.
D = Diameter in sixteenths of an inch.

(22.) $W = \dfrac{D^2}{9} \therefore D = \sqrt{9W}.$

In this district, the factor of safety used for cage chains is probably about 10, not 6.

Cast Iron Pipes.
Th = Thickness of metal in inches.
D = Diameter of pipe in inches.
H = Head of water in feet that will burst pipe.

(23.) $H = \dfrac{72{,}000 \, Th}{D} \therefore Th = \dfrac{DH}{72{,}000}.$

If W = Weight per linear foot of cast iron pipes in lbs.
D = Outside diameter in inches.
d = Inside do. do.

(24.) $W = 2\cdot 45 \, (D^2 - d^2).$

The weight of the two flanges may be taken as equal to one foot of pipe.

Boiler Shells.
Th = Thickness of plate in inches.
D = Diameter of boiler in inches.
P = Bursting pressure of steam in lbs. per square inch.

(a) *Iron Boilers.*

(25.) $P = \dfrac{50,000 \, Th}{D} \therefore Th = \dfrac{PD}{50,000}$ (single riveted).

(26.) $P = \dfrac{60,000 \, Th}{D}$ and $Th = \dfrac{PD}{60,000}$ (double riveted)

(b) *Steel Boilers.*

(27.) $P = \dfrac{70,000 \, Th}{D} \therefore Th = \dfrac{PD}{70,000}$ (single riveted).

(28.) $P = \dfrac{90,000 \, Th}{D} \therefore Th = \dfrac{PD}{90,000}$ (double riveted).

Though 6 is usually considered sufficient as the factor of safety, 8 agrees better with the practice of this district.

Boiler Tubes (Iron).
   Th = Thickness of plate in inches.
   D = Diameter of tube in inches.
   L = Length of tube in inches.
   P = Collapsing pressure in lbs. per square inch.

(29.) $P = \dfrac{9,672,000 \, Th^2}{LD} \therefore Th = \sqrt{\dfrac{PLD}{9,672,000}}$.

Masonry Pillars.
   W = Crushing load in tons.
   A = Area of red brick pillar in square inches.

(30.) $W = \cdot 38 A \therefore A = \dfrac{W}{\cdot 38}$

See also T in Table IX.

Beams.
   L = Length of beam or span in inches.
   B = Breadth of beam in inches.
   D = Depth of beam in inches.
   W = Breaking load in tons.
   K = Coefficient of rupture. (See Table VIII.)

(31.) $W = \dfrac{KBD^2}{L}$ when one end is fixed, and the other end loaded.

(32.) $W = \dfrac{2KBD^2}{L}$ when one end is fixed, and the load distributed.

(33.) $W = \dfrac{4KBD^2}{L}$ when both ends are supported, and the load is in the centre.

(34.) $W = \dfrac{6KBD^2}{L}$ when both ends are fixed, and the load is in the centre.

(35.) $W = \dfrac{8KBD^2}{L}$ when both ends are supported, and the load distributed.

(36.) $W = \dfrac{12KBD^2}{L}$ when both ends are fixed, and the load is distributed.

Circular beams with radius R inches; substitute $4\cdot 7 R^3$ for BD^2 in the above formulæ.

TABLE VIII.
Values of K for Different Materials.

Wrought Iron	$K = 3\cdot 40$
Cast Iron	$K = 2\cdot 30$
English Ash	$K = 0\cdot 95$
Beech	$K = 0\cdot 65$
Fir (Spruce)	$K = 0\cdot 60$
English Oak	$K = 0\cdot 75$
African Oak	$K = 1\cdot 10$
Red Pine	$K = 0\cdot 65$
Yellow Pine	$K = 0\cdot 50$
Memel Pine	$K = 0\cdot 60$
Pitch Pine	$K = 0\cdot 75$

For most purposes, the breadth and depth of the beam should be proportioned, so that, in round numbers, the depth be about $1\tfrac{1}{2}$ times the breadth.

If then we put $D = 1\cdot 5 B$, (31) becomes

(37.) $W = \dfrac{KB(1\cdot 5B)^2}{L} \therefore B = \sqrt[3]{\dfrac{WL}{2\cdot 25 K}}$

and similarly for the rest.

Molesworth (pp. 119—120, 19th ed.) gives for cast and wrought iron girders :—

Cast-Iron Girders.

D = Depth of girder in inches, including flanges.
A = Area of bottom flange in inches (*i.e.*, width × thickness).
S = Span in inches.
W = Breaking weight in tons.

Supported at both ends with load :—

(38.) On centre, $W = \dfrac{25\,AD}{S} \therefore D = \dfrac{WS}{25A}$

(39.) Distributed, $W = \dfrac{50\,AD}{S} \therefore D = \dfrac{WS}{50A}$

Area of top flange if the load is applied on the top $= \dfrac{A}{3}$

If applied on the bottom flange $= \dfrac{A}{2}$. And $D = \dfrac{S}{12}$ (about).

Wrought-Iron Plate Girders.

L = Span in feet.
W = Weight distributed in tons.
D = Effective depth of girder in feet.
S = Strain on top and bottom flange at centre in tons.

(40.) $S = \dfrac{WL}{8D} \therefore D = \dfrac{WL}{8S}$

In compression, iron may be strained 4 tons ; in tension, 5 tons per square inch.

Long Square Columns—length more than 30 times diameter.

W = Breaking load in tons.
B = Breadth in inches.
L = Length in feet.
K = Coefficient of rupture.

(41.) $W = \dfrac{KB^4}{L^2} \therefore B = \sqrt[4]{\dfrac{WL^2}{K}}$

K for dry Memel = 7·81.
K for dry Oak = 10·95.

If in a damp situation as pulley frames, K must be taken rather less, say 6 and 9 respectively.

Dams, Tubbing, &c.

k = Thickness in inches.
r = External radius in inches.
T = Ultimate crushing strength in lbs. per square inch (See Table IX.).
p = Head of water in lbs. per square inch.

Cylindrical Dam, Walling, or Tubbing.

(42.) $k = r \left\{ 1 - \sqrt{1 - \dfrac{2 0 p}{T}} \right\}.$

Spherical Dam.

(43.) $k = r \left\{ 1 - \sqrt[3]{1 - \dfrac{1 5 p}{T}} \right\}.$

10 is taken as the factor of safety, and is allowed for in the formulæ.

See "Internal Stress in Cylindrical and Spherical Dams," by W. Steadman Aldis. Trans. N.E.I., xxxii.

TABLE IX.

Wrought-Iron	T = 38,080
Cast-Iron	T = 107,520
Beech	T = 8,500
Oak	T = 10,000
Pitch Pine	T = 6,500 (?)
Brick ord. red	T = 800
Do. Stourbridge fire	T = 1,717
Sandstone	T = 2,185 to 7,884
Concrete	about T = 2,000

(Molesworth.)

Another Formula for Cast-Iron Tubbing.

x = Required thickness in feet.
P = Vertical depth in feet.
D = Diameter of pit in feet.

(44.) $$x = \cdot 03 + \frac{PD}{50,000}$$

(Greenwell's "Mine Engineering.")

The following books may be consulted :—Barlow's "Strength of Materials"; Box's "Strength of Materials"; "Materials and Construction," by Campin, Weale's Series; and Molesworth's "Pocket Book of Engineering Formulæ."

TIMBER.

The most suitable timber for pit props, baulks, &c., is fir and pine, because—though weaker and less durable than oak, elm and some other timber—it is light, cheap, straight, and elastic.

The mining timber used in the North of England comes chiefly from Sweden, and is the product of the Scotch fir (*Pinus sylvestris*) and the spruce (*Abies communis*), known to the trade as "red and white wood" respectively. A good deal of native-grown larch (*Abies larix*) too is used, more especially for sleepers.

Beech, grown in the district, is used for nogs. Pitch pine (*Pinus rigida*) from North America, for pump spears, pulley frames, &c.

Timber Measure.

Props are bought and sold per 72 running feet, the price depending upon the diameter. Larger sizes per cubic foot, per load, or per standard. To calculate the number of cubic feet in round timber: gird the log round the middle with a string, and one fourth of the girth squared × the length = cubic content. If the log be very irregular, divide it into several lengths and measure each separately.

    1 load = 40 cubic feet, unhewn timber.
      ,,   = 50   ,,   squared ,,
      ,,   = 600 superficial feet, in 1 inch deals or planks.
      ,,   = 400   ,, $1\frac{1}{2}$,,
and so on, equal in each case to 50 cubic feet.

    1 square of flooring = 100 superficial feet.

Battens are 7 inches wide, deals 9 inches, and planks 11 inches.

1 standard	= 165	cubic feet square timber.
,,	= 150	,,	partly squared.
,,	= 100	,,	round.
,,	= 1500	running feet of 3 inch.
,,	= 1200	,,	4 ,,
,,	= 1000	,,	5 ,,

A standard of timber occupies about the same space on board ship as $3\frac{3}{4}$ tons of coal; but this partly depends upon the shape of the ship.

Cost of Timbering.

This is very variable, depending upon the conditions of the mine and the price of timber.

Callon cites Grand-Combe $3\frac{1}{2}d.$ per ton of coal. Evrard, a colliery in the Pas-de-Calais, 10·088$d.$; and one in Belgium, 1$s.$ 3$d.$ Drinker, quoting Rhiza, gives the consumption of 54 German mines from 1·35 to 8·58 cubic feet of timber per 100 cubic feet of coal. Average, 3·40 cubic feet.

At a colliery in Durham, using larch, the cost, in 1877, was 4·08$d.$ per ton of coal worked. And during the same year, at a colliery in Northumberland, using Norway, the cost was 4·77$d.$

The following books may be consulted:—Templeton's "Workshop Companion"; Laslett's "Timber and Timber Trees"; Rattray and Mills' "Forestry and Forest Products."

EXPLOSIVES.

General rule 12 of the Mines Act, 1887, deals with explosives. The chief points are as follows :—No explosive may be taken into the mine except in cartridges in a canister containing not more than 5lbs. Charging tools of iron or steel are forbidden, and coal may not be used for tamping. No explosive may be pressed into an insufficient hole. A charge may not be unrammed, and no hole may be bored for a charge at less than six inches from a hole where the charge has missed fire. In any place where a safety lamp is required, or which is dry and dusty, shots may only be fired by a person appointed for the purpose, who must first examine all places within 20 yards. If gas has been reported a shot may not be fired unless there is not sufficient gas at or near the place of firing to render it unsafe; or unless the explosive be of such a nature that it cannot fire gas. If the place be dry and dusty a shot may not be fired unless the place to a distance of 20 yards be watered; or, should watering damage the roof or floor, unless the explosive be of such a nature that it cannot inflame gas or dust. If the place be dry and dusty and be on, or contiguous to, a main haulage road, a shot may not be fired unless both the last-mentioned conditions have been observed; or else such one of them as may be applicable, and also all workmen have been removed from the seam and from all seams communicating with the shaft upon the same level, except the shot firers and such others not exceeding ten as are necessarily employed in attending to furnaces, machinery, signals, horses and inspection.

The Explosives Act, 1875, is too long to abstract; but a few points may be mentioned. No explosive may be kept for sale without a licence. Not more than 20 lbs. of gunpowder, and 150 lbs. of safety cartridges (as ordinary shots) or 15 lbs. of any other explosive, or in lieu of any less amount of gunpowder not so kept, half that amount of other explosive may be kept for private use without a licence. Cartridges for blasting may not be made in a private house; they must be bought ready made, or manufactured in a workshop in connection with, but detached from, (25 to 100 yards) the store. The store must not be situated in a mine or quarry where persons are employed ; or within a certain distance (the exact distance depending upon the quantity of explosive for which it is licensed ; but 200 yards is the maximum, and should houses, etc., be subsequently built within the prescribed zone, the store must be removed), of houses, workshops, railways, roads, fires, etc. It must be substantially built of brick, stone, or concrete ; or be excavated in solid rock, earth, or mine refuse not liable to ignition ; and so made and closed as to prevent unauthorised persons from having access. There must be no exposed iron, steel, or grit in the building. Nothing may be kept in the store but the explosive and the necessary implements, which must be made of copper, wood, or brass. Lightning conductor required, unless the store be underground or licensed for less than 1,000 lbs. of gunpowder. No person under 16 to enter, except under supervision of a grown-up person. The quantity of gunpowder that may be kept varies from 300 to 4,000 lbs., according to the character of the store. If licensed for mixed explosives, 300 to 4,000 lbs. of powder, and, in addition, 1,500 to 20,000 lbs. of safety cartridges (the cases are included in the weight) ; in lieu of each lb. of powder, $\frac{1}{2}$ lb. of any other explosive may be kept ; and, in addition to each lb. of powder, 5 lbs. of safety fuzes. A copy of the rules must be affixed to the store. There are several common sense regulations, such as no smoking allowed, etc. See " Guide Book to the Explosives Act," by Major Majendie.

All explosives exert an equal force in every direction.

An explosive takes effect along the line of least resistance. In a homogeneous material this will be the shortest line from the charge to the face, and will be most efficient when at right angles to the shot-hole, and least efficient when it coincides with the axis of the shot-hole. The quantity required varies as the cube of the line of least resistance. A cartridge, one inch diameter, and 38 inches long, contains one lb. of powder. The chemical changes that take place when powder is fired, may be roughly represented as follows :—

(45.) $S + 3C + 2KNO_3 = 3CO_2 + N_2 + K_2S$.

Gunpowder fires at about 482° Fahr. and expands to at least 1,500 times its original volume.

(Miller's " Inorganic Chem.")

TABLE X.

Explosive.	Composition.	Heat Evolved	Vol. of Gas per lb.	Product indicating Blasting Effect.
Gunpowder.	Potassium Nitrate......74·70 Sulphur..................12·45 Charcoal ....12·25	1,093	3·61	3,945
Nitro-glycerine.	$C_3H_5O_3(NO_2)$	2,372	11·41	27,064
Dynamite.*	Nitro-glycerine.........75 Silica 25	1,780	8·56	15,236

(Proc. I.C.E., XLIII.)

Tons of coal got in 1875 per lb. of powder employed both in hewing and stone work—

    Northumberland Steam Coal . . . 5·95
    Northumberland House, Coking Coal, &c . 25·80
    Durham House and Gas Coal . . . 21·00
    Durham Coking Coal, &c. . . . 23·99
    Cumberland . . . . . . 21·42

    Whole district . . . 9·97

(J. B. Atkinson.)

In Pennsylvania in 1881 and 1882, 2·23 tons only; but the seams lie at high angles, and there is much stone work.

Substitutes for Explosives.

The wedge, wedge and feather, Macdermott's multiple wedge and feather, Macdermott's screw wedge, Macnab's

---

* Dynamite, if used with water tamping, will get coal in good condition; and has this advantage over gunpowder used with water tamping, that a blown out shot will not fire gas. (See Report of Accidents in Mines Commission, 1886.)

hydraulic cartridge, the Haswell Mechanical Coal-getter, the Seaton Delaval Detacher, and many others. See Trans. N.E.I., ii., xii., xiv., xix., xxiii., xxxiii.

Lime Cartridges.

Quick lime + water in excess = slaked lime + steam.

(46.) $CaO + H_2O + Aq = CaH_2O_2 + Aq.$

Mountain limestone is calcined, ground to a fine powder, and formed into cartridges $2\frac{1}{2}$ inches diameter, with a groove along the side by means of an hydraulic pressure of forty tons. The shot-holes being drilled, an iron tube half-an-inch in diameter, having a small external groove on the upper side, and provided with perforations, is inserted along the whole length of the bore-hole. This tube is enclosed in a bag of calico, covering the perforations at one end, and has a tap at the other. The cartridges are then inserted and tamped. A force pump is connected with the tap by means of a flexible pipe and water, equal in bulk to the quantity of lime used, is forced in. The water being driven to the far end of the shot hole through the tube, escapes along the groove, and through the perforations and the calico, flowing towards the tamping into the lime and driving out the air before it. The tap is then closed. The pressure of steam generated by the usual charge of seven cartridges is 2,850 lbs., and the expansion of the cartridge about five times its original size.

The advantages claimed for this system are :—Absolute immunity from explosion of gas, there being no fire or flame. There is no smoke or noxious smell. The roof is not shaken, and the coals in falling produce less dust. Skilled labour is unnecessary, and the coal can be got with much less exertion than by wedging. Major Paget Mosley. Trans. Midland Inst. of Mining, Civil, and Mechanical Engineers. Vol. VIII., pp. 87—93.

It has been averred that the heat given off is sufficient to ignite gas, but Abel says that the maximum heat produced by the slaking of the lime is 700 degrees, whereas it takes a temperature of 2,000 degrees to ignite gas. On the other hand, coal dust ignites at a temperature much below 700 degrees Fah.

MACHINERY.

Nature and Uses of Machinery.

The use of machinery is to transmit and modify motion and force. In the action of a machine, the three following things take place :—1st. Some natural *source of energy* communicates *motion* and *force* to a part of the mechanism called the *prime mover*. 2nd. The motion and force are transmitted from the prime mover through the train of mechanism to the *working piece ;* and, during that transmission, the motion and force are modified in amount and in direction, so as to be rendered suitable for the purpose to which they are applied ; and 3rd, The working piece, by means of its motion, or of its motion and force combined, accomplishes some useful purpose. (Rankine, M.M., p. 1.)

Some of these terms require explanation, viz. :—

Force may be defined as an action between two bodies, causing or tending to cause, rest or motion. The British unit of force is the force required to support a weight of one lb. at London, or roughly, at any other place on the globe. *Work* may be defined as the combination of force and motion. The unit of work is a force of one lb. exerted through a distance of one foot. *Power* may be defined as the speed of doing work. The unit of power is a force of one lb. exerted through a distance of one foot in one minute.

A horse-power is equal to 33,000 of the above units of power.

It is sometimes convenient to use the second, or the hour as the unit of time instead of the minute. One British horse power then is :—550 foot lbs. per second = 33,000 foot lbs. per minute = 1,980,000 foot lbs. per hour.

We have then the following rules :—

(47.) Units of Work = Force in lbs. × Distance in feet.

(48.) Units of Power = $\dfrac{\text{Force in lbs.} \times \text{Distance in feet.}}{\text{Time in minutes.}}$

(49.) Units of Horse-power = $\dfrac{\text{Force in lbs.} \times \text{Distance in feet.}}{\text{Time in minutes} \times 33{,}000.}$

Energy may be defined as *the power of doing work.*

Heat is a form of energy. The unit of heat (or thermal unit) is, approximately, the quantity of heat required to raise one lb. of water one degree Fahr. Different bodies require very different quantities of heat to effect in them the same change of temperature, and the quantity of heat that one lb. of a body requires to raise its temperature one degree is called the specific heat of that body.

TABLE XI.
Specific Heats.

			At constant pressure.	At constant volume.
Water............1·0000	Air ....0·2379		0·1686	
Cast Iron ......0·12983	Carbonic Acid......0·2164		0·1711	
Wrought Iron ...0·11379	Carbonic Oxide ...0·2479		0·1763	
Copper ....0·09515	Vapour of Water ...0·4750		0·3640	
Silver............0·05701	Nitrogen ....0·2440		0·1727	
Tin..............0·05695	Sul. Hydrogen ...0·2423		0·1833	
Gold ....0·03244	Hydrogen ....3·4046		2·4046	
Lead ....0·03140	After-damp......0·268		0·196	
Coal ....0·2777	Oxygen ....0·2182		0·1555	
Coke ....0·20085				
Calcium ....0·1670				
Slaked Lime ...0·223				

(Regnault.)

We have, then, the following rule :—

The units of heat required to raise a given body, a given number of degrees = the weight of the body × the number of degrees × the specific heat of the body.

If, therefore, U = Units of heat.
                D = Degrees Fahr. the body is heated.
                S = Specific heat of body. (See Table XI.)
                W = Weight of body in lbs.

(50.) $U = WDS \therefore D = \dfrac{U}{WS}.$

D

Connexion between heat and work.—One unit of heat = 772 units of work. A lb. of coal yields about 14,000 units of heat. An engine, therefore, consuming one lb. of coal per hour, should develop $\frac{14,000 \times 772}{33,000 \times 60} = 5\frac{1}{2}$ horse-power nearly.

One lb. of marsh gas yields about 23,550 units of heat. One lb. of hydrogen about 62,000. One lb. of illuminating gas about 22,000. Electricity is a form of energy, and may be converted into heat as in the electric light, or into work as in an electric engine. One lb. of zinc reacted upon by sulphuric acid in a battery yields 1,018 units of work.

Energy is indestructible, but in converting one form of energy into another, there is always practically great waste. For example: One lb. of coal, though it yields 14,000 units of heat, and should, therefore, give us $5\frac{1}{2}$ horse-power, will only give, in the best steam-engines, about $\frac{1}{2}$ of one horse-power.

The principal sources of energy are :—Food, fuel, heads of water, and the wind.

The principal prime movers are :—Men and horses, steam-engines, water-wheels, and wind-mills.

The train of mechanism which connects the prime mover with the working piece may consist of wheels, levers, spears, ropes, a fluid, electricity, etc.

The working piece may be a bit, pump, cage, tub, etc.

We see then that *machinery* enables us to make use of the *energy* Nature provides.

The following books may be consulted:—The Conservation of Energy, Balfour Stewart; Energy in Nature, Carpenter; Heat a Mode of Motion, Tyndall.

A. Men and Horses.

Food is the source from which men and horses obtain their energy; their efficiency is very great as compared with the efficiency of a steam-engine, about 27 per cent. of the units of heat yielded by a horse's food being turned into mechanical energy or work, against about 10 per cent. in a steam-engine.

Mr. Hunting is the great authority in the North of England on the feeding and management of colliery horses, and his views are embodied in a paper published in the Trans. N.E.I., vol. xxxii. Table XII., extracted from this paper, shews that beans and peas contain the largest pro-

TABLE XII.

	Water.	Woody Fibre.	Starch, Gum, Sugar, & Fat.	Nitrogenous matter.	Ash or Saline.	Total.	Remarks.
Beans or Peas...	14·5	10·0	46·0	26·0	3·5	100·00	Constipating.
Barley......	13·2	13·7	56·8	13·0	3·3	100·00	Not more than 25% of the total corn mixture.
Oats.........	11·8	20·8	52·0	12·5	3·0	100·10	Form with hay a good food; but costly.
Maize	13·5	5·0	67·8	12·29	1·24	99·83	Laxative.
Hay.........	14·0	34·0	43·0	5·0	5·0	101·00	50 lbs. old land hay = 60 lbs. of new land hay.
Carrots ...	85·7	3·0	9·0	1·5	0·8	100·00	
Bran	...	...	...	...	...	...	Laxative, and useful as a bulky palatable article, but has little feeding value.
Linseed ...	...	...	...	...	...	...	Laxative.

portion of nitrogenous matter; whilst oats, maize, and barley are of equal feeding value per stone. Beans or peas, being constipating, must be mixed with maize or bran. The quantity of food must be regulated by the amount of work; but about 100 lbs. of mixed corn, crushed and mixed with about 56 lbs. of chopped hay, will form an average week's provender for each horse. He considers that beans (or peas) and hay with either, 1, Oats and bran; 2, Barley and bran; 3, Oats and maize; 4, Maize—will form equally good mixtures, and we must be guided in our selection by price.

A horse should be put down the pit at about 5 to 7 years of age; should travel from 14 to 16 miles per day, in a fairly level mine at a walking pace, and last about 7 years. One horse-keeper is required for 12 horses, or for 16 ponies. A pony should not be put into the pit under 3 years of age.

The cost of feeding, etc., is very variable; but we may say, roughly, that to maintain a stud of 100 horses (or their equivalent in ponies) we must expend yearly:—

	£	s.	d.
Provender @ 10/- per horse per week	2,600	0	0
Veterinary surgeon	150	0	0
Head horse-keeper @ 25/- per week	65	0	0
8 assistant ,, @ 20/- ,,	416	0	0
2 loftmen @ 18/- per week	93	12	0
Shoeing (labour and materials) ...	120	0	0
Saddlery (,,) ...	210	0	0
Drugs	10	0	0
Renewals, less sales	350	0	0
Candles, brushes, clipping, etc. (say)	50	8	0
Total	£4,065	0	0

equal to £40 13s. 0d. per horse per annum.

But each horse will require a boy to drive it; and in the North of England the horse-keepers and loftmen will receive free houses and fire-coal. So that the total cost of each horse will amount to nearly £60 a year.

The following weights and measures are used:—

14 lbs. = 1 stone; 4 pecks = 1 bushel; 8 bushels = 1 quarter.

A bushel of oats should weigh not less than 42 lbs.
," barley ," 56 ,,
," maize ," 60 ,,
," beans ," 63 ,,
," peas ," 63 ,,

From Table XIV. we see that a draught horse, under conditions somewhat similar to those which are found in mines, can do $120 \times 3.6 \times 60 = 25,920$ foot lbs. per minute during a shift of eight hours. This is rather less than the power of a pit horse, as calculated from experiment by Mr. Nicholas Wood, Trans. N.E.I., iii.

TABLE XIII.
WORK OF A MAN AGAINST KNOWN RESISTANCES.

Kind of Exertion.	R. lbs.	V. ft. per sec.	$\frac{T''}{3600}$ hours per day.	R.V. ft.-lbs. per sec.	R.V.T. ft.-lbs. per day.
1. Raising his own weight up stair or ladder.........	143	0.5	8	72.5	2,088,000
2. Hauling up weights with rope, and lowering the rope unloaded	40	0.75	6	30	648,000
3. Lifting weights by hand..	44	0.55	6	24.2	522,720
4. Carrying weights up stairs, and returning unloaded.	143	0.13	6	18.5	399,600
5. Shovelling up earth to a height of 5 ft. 3 in.	6	1.3	10	7.8	280,800
6. Wheeling earth in barrow up slope of 1 in 12, ½ horiz. veloc. 0.9 per ft. sec. and returning unloaded	132	0.075	10	9.9	356,400
7. Pushing or pulling horizontally (capstan or bar)	26.5	2.0	8	53	1,526,400
8. Turning a crank or winch	12.5	5.0	?	62.5	...
	18.0	2.5	8	45	1,296,000
	20.0	14.4	2 min.	288	...
9. Working pump	13.2	2.5	10	33	1,188,000

TABLE XIV.
Work of a Horse against known Resistances.
(Three 14-hand ponies = two horses; and two small ponies = one horse.)

Kind of Exertion.	R.	V.	$\frac{T''}{3600}$	R.V.	R.V.T.
1. Cantering and trotting drawing a light railway carriage (thorough-bred)	min. 22½ mean 30½ max. 50	14⅔	4	447½	6,444,000
2. Horse drawing cart or boat walking (draught horse)	120	3·6	8	432	12,441,600
3. Horse drawing a gin or mill walking	100	3·0	8	300	8,640,000
4. Ditto trotting	66	6·5	4½	429	6,950,000

Explanation.—R., resistance; V., effective velocity = distance through which R is overcome ÷ total time occupied, including the time of moving unloaded, if any; T'', time of working, in seconds per day; $\frac{T''}{3600}$, same time, in hours per day; R.V., effective power in foot-pounds per second; R.V.T., daily work in foot-pounds.

The weight of the man or horse is not included, except in No. 1 of the first Table. (Rankine, S. E.)

The annual cost of a pit horse, including depreciation and driver's wages, is about £60.

B. Hydraulic Motors.

A head of water is the source from which hydraulic motors receive their energy; they are the most efficient of all prime movers, utilising about 80 per cent. of the units of work stored up in the head of water, as compared with 27 per cent. and 10 per cent. utilised by horses and steam-engines.

A head of water can be made use of in one or other of the following ways, viz.:—

1st. By its weight, as in the water-balance, and overshot wheel.

2nd. By its pressure, as in the hydraulic engine.

3rd. By its impulse, as in the undershot wheel.
4th. By a combination of the above.

Bearing in mind our definition of the unit of power, see p. 32, we see that the maximum units of power a head of water can yield are equal to the volume in cubic feet falling per minute × pressure in lbs. per square foot, *i.e.*, the volume, × the weight in lbs. per square foot of the column or head of water.

This general statement is true for all fluids—water, steam, air, etc.—and may be expressed as follows :—

If V=Volume of any discharged fluid in cubic feet.
P=Effective Pressure at which it is discharged in lbs. per square foot.
U=Units of work in foot lbs. that can be got from the fluid's discharge without expansion.

(51.) U=PV.

In speaking of water, however, it is generally more convenient to say :—

If W=Weight of water falling per minute in lbs.
H=Head of water in feet.
U=Units of work in foot-lbs. per minute.

(52.) U=WH.

The amount of rainfall is an important factor that must be taken into account in calculating the water-power of any place. The average annual rainfall in different parts of Britain ranges from 22 inches to 140 inches. At Newcastle it is from 26 to 27 inches.

(53.) One inch in depth per acre weighs 101 tons.

C. Wind-mills.

The wind is the source from which wind-mills derive their energy; their efficiency is about 29 per cent. (Smeaton.)

The power of a wind-mill might be calculated from the rule already given, viz. :—(51.) U=PV.

In which U=units of work in foot-lbs. per minute.
P=pressure of the wind in lbs. per sq. foot.
V=volume in cubic feet of the cylinder of wind passing the sails each minute, the diameter of the cylinder being equal to the diameter of the sails.

But that, the sails being in motion, the value of P cannot easily be obtained. If the sails were stationary, the wind-pressure could be found as follows:—

Let V_1=velocity of the wind in miles per hour.
P=pressure in lbs. per sq. foot.

(54.) $P = \dfrac{V_1^2}{200}$.

In these circumstances, the following rule should be used:—

U=units of work in foot lbs. per sec.
W=weight in lbs. of the cylinder of wind passing the sails each second, the diameter of the cylinder being equal to the diameter of the sails.
V=velocity of wind in feet per sec.
HP=horse-power.

(55.) $U = \dfrac{WV^2}{64}$.

(56.) $HP = \dfrac{WV^2}{64 \times 550}$.

(57.) Effective horse-power $= \dfrac{0\cdot 29 WV^2}{64 \times 550}$.

The average velocity of the wind in England is about 13 miles an hour.

D. The Steam Engine.

Fuel is the source from which the steam-engine derives its energy. One lb. of very good coal yields about 14,000 units of heat = 14,000 × 772 = 10,808,000 units of work. The best steam-engines consume about 2 lbs. of coal per effective HP per hour = 14,000 × 2 × 772 = 21,616,000 units of work; but one horse-power per hour is only 33,000 × 60 = 1,980,000 units of work. The efficiency, therefore, of the best engines is only :—

$$(58.) \quad \frac{1,980,000 \times 100}{21,616,000} = 9\cdot15 \text{ per cent.}$$ of the total energy stored up in the coal.

Why is this so?—The loss is in part due to heat uselessly expended in warming the sides of the boiler, steam-pipes, cylinders, etc., and air in contact with them; also, to imperfections in the machinery. These may be in part removed with improvements in machinery. For example, the following Table, quoted from *Coal: its History and Uses*, p. 227, illustrates the large saving of heat effected in the past by improvements in the steam-engine.

TABLE XV.
100 UNITS OF HEAT SUPPLIED.

Description of Engine.	Heat transmitted to Boiler.	Heat used in doing Work.
Unjacketed	54	5·9
Jacketed	54	6·6
Steam superheated	60	8·7
Working with Siemen's Regenerator	60	10·0
Engines of H.M.S. "Briton"	—	11·1

But the nature and properties of steam account for a very large proportion of the loss, and this cannot be recovered.

It has been estimated that 20 per cent. of the energy stored in the coal passes up the chimney, 10 per cent. is lost by radiation, 10 per cent. is turned into work, and the remaining 60 per cent. is wasted.

If T_1 = Absolute temperature of steam on admission, *i.e.*, temp. Fahr. + 459°.

T_2 = Absolute temperature of exhaust steam, *i.e.*, temp. Fahr. + 459°.

E = Maximum efficiency of a theoretically perfect steam-engine.

(59.) $E = \dfrac{T_1 - T_2}{T_1}$.

The Actual or Indicated horse-power is the measure of the capacity of the engine for doing work :—

If, therefore, A = Net area of piston in square inches.
     P = Effective pressure of steam in lbs. per square inch.
     S = Mean speed of piston in feet per minute.
     HP = Horse-power.

(60.) $HP = \dfrac{APS}{33,000} \therefore A = \dfrac{33,000\,HP}{PS}$.

In a *high pressure non-condensing engine* P = the pressure of the steam in the boiler, less the pressure of the atmosphere, *i.e.*, P = the reading of the steam-gauge.

The low-pressure condensing-engine.—The actual horse-power is given by formula (60), but P now is equal to the pressure of the steam in the boiler, less the pressure of the vapour in the condenser, *i.e.*, P = reading of the steam-gauge in lbs. + half the reading of the vacuum gauge in inches.

Engines working with expansion.—The actual horse-power is given by formula (60), but in non-condensing engines working with expansion, P is given by the following rule :—

MP = Mean pressure.
PB = Boiler pressure = reading of steam-gauge + 15 lbs.
L = Total length of stroke.
Q = Length of stroke before cut off.

(61.) $MP = \dfrac{PBQ}{L}\left(1 + \text{Hyp. log.}\dfrac{L}{Q}\right)$.

Hyp. log. $\dfrac{L}{Q}$ will be found in the Table, see page 132.

Finally P = MP − 15. And in condensing engines working with expansion P is equal to MP as obtained from Formula (61), less the pressure of the vapour in the condenser.

The compound engine prevents shocks and cooling. See table, pages 50 and 51, for the fall in temperature corresponding with any given fall in pressure. To calculate the horse-power of a compound engine, treat it as if it were a one-cylinder engine, of the size of the large cylinder of the compound engine, and use the preceding formulæ (neglecting loss due to steam passages).

By means of the above formulæ, the size of engine required to do any given work can be calculated, the pressure of steam and speed of piston being known. In the case of a compound engine, find the size of a single cylinder engine required to do the work, and add to it a small cylinder proportioned as follows :—

If A = Area of piston of large cylinder.
a = Area of piston of small cylinder.
R = Ratio of expansion.

(62.) $$a = \frac{A}{\sqrt{R}}$$

Half the total expansion should be carried out in each cylinder.

The Indicator is used for finding the horse-power when greater accuracy is required. It also enables the Engineer to find out and localise any defects in the working of an engine.

Nominal horse-power is an expression used for commercial purposes only, and gives merely the size of the engine.

If D = Diameter of cylinder in inches.
NHP = Nominal horse-power.

(63.) $NHP = \dfrac{D^2}{10}$ (High pressure or non-condensing engine where $P = 21$ and $S = 200$.)

(64). $NHP = \dfrac{D^2}{30}$ (Low pressure or condensing engine where $P = 7$ and $S = 200$.)

The effective horse-power, or the useful work done, is a very variable quantity, depending upon the efficiency of the engine, the way in which it is connected with its work, the nature of the work, etc., etc. Speaking roughly, the effective horse-power of the engines used at our mines is about one half of their actual horse-power.

Proportions of engines.—See Molesworth, p. 325, xix. Ed.

If A = Transverse sectional area of a cylinder.
   D = Diameter of a cylinder.
   C = Circumference of a cylinder.
   V = Volume of a cylinder.
   L = Length of stroke.

(65.) $A = \cdot 7854\, D^2 \therefore D = \sqrt{\dfrac{A}{\cdot 7854}}$

(66.) $C = 3 \cdot 1416\, D \therefore D = \dfrac{C}{3 \cdot 1416}$.

(67.) $V = \cdot 7854\, D^2 L$.

The cost for the repairs of colliery engines has been estimated at 5*s.* per horse-power per annum. See Trans. N.E.I. xvii.

BOILERS.

Strength of Boilers.

Steam expands equally in all directions; therefore the pressure is equal in all directions. It is reasonable, therefore, to suppose that the strongest shape for a boiler is that figure which is similar to itself in all directions, viz., a sphere. As, however, a spherical boiler would be inconvenient, a cylindrical boiler, with hemispherical ends, is the strongest practical form.

1st. *Strength to resist internal pressure.*—Let a cylindrical boiler be L inches long, D inches diameter, and let P = pressure of steam in lbs. per sq. in. Let T = the tensile strain in lbs. per linear inch to which the boiler plates are subjected.

Then the total force tending to tear asunder the sides (*i.e.*, to open the horizontal seams), is equal to the pressure of the steam in lbs. per sq. in. × area in sq. inches over which it is exerted = PDL. And this is resisted by the tensile strength of the side plates, which are 2L inches in length. The tensile strain upon each linear inch of the side plates (*i.e.*, upon the horizontal seams) is, therefore,

$$(68.)\ T = \frac{PDL}{2L} = \frac{PD}{2}.$$

In the same way, the strain per linear inch upon the end plates (*i.e.*, upon the vertical seams) is

$$(69.)\ T = \frac{\cdot 7854 PD^2}{3 \cdot 1416 D} = \frac{PD}{4}.$$

We see, then, that the horizontal seams are subjected to double the strain upon the vertical seams. For this reason, boiler-plates are sometimes rolled in rings, so that the boiler may have no horizontal seams at all.

The ultimate tensile strength of ordinary iron boiler plate is 20 tons per sq. inch; which is reduced by single riveting to about 25,000 lbs., and by double riveting, to about 30,000 lbs. We deduce from these facts, and from formula (68), the formulæ (25) and (26).

The use of steel plates for colliery boilers is hardly yet established. We may, however, take their ultimate tensile strength at 30 tons per sq. inch; and the reduction of strength due to riveting the same as in iron plates. From these data formulæ (27) and (28) are calculated.

2nd. *Strength to resist external pressure.*—The flues are subjected to external pressure. See formula (29). Where they are strengthened by rings, L = length between rings.

The weight of a boiler may be calculated from the following rule, viz.:—Weight in lbs. = surface in sq. ft. × 6 for plates $\frac{1}{8}$-inch.

3rd. *Practical remarks on iron boilers.*—Generally in the North, $\frac{3}{8}''$ plates are used for sides and hemispherical ends; rivets, $\frac{3}{4}'' \times 1\frac{3}{4}''$, placed 2" apart, $1\frac{1}{2}''$ lap. For flat ends, $\frac{5}{8}''$ to $\frac{6}{8}''$ plates strengthened with gusset stays or rods. Tubes, $\frac{3}{8}''$ to $\frac{4}{8}''$. Diameter of low pressure boilers, 6 to 10 ft.; high pressure, 4 to 6 ft.

Boiler Fittings.

Man-hole, weighted safety-valve, spring safety-valve, steam-gauge, float, and glass water-gauge, sludge steam and feed-pipes, blow-off cock, damper, etc.

The steam-gauge gives the pressure of the steam in the boiler, less the pressure of the atmosphere.

The lever safety-valve :—

P = Pressure of steam in lbs. per square in., less atmospheric pressure.
L = Length of lever from fulcrum to weight in inches.
W = Weight of weight in lbs.
W' = Weight of lever in lbs.
W'' = Weight of valve in lbs.
D = Distance of the centre of gravity of the lever from the fulcrum in inches.
A = Area of valve in square inches.
I = Length of lever between fulcrum and valve in inches

$$(70.)\ P = \frac{\dfrac{WL}{I} + \dfrac{W'D}{I} + W''}{A}$$

Explosions.

Explosions may be produced by—1st. Defective, or neglected, or tampered-with safety-valves, producing excessive pressure. 2nd. Defective water supply. 3rd. Incrustation, producing over-heating. 4th. Corrosion, weakening the boiler.

The remedies for 1st and 2nd are obvious.

3rd. *Incrustation* (as distinguished from mere sediments due to dirty water, which are easily blown out, or gathered up, by means of sediment collectors) is due to the presence of salts in the feed water (carbonates and sulphates of lime and magnesia for the most part), which are precipitated when the water is heated, and form hard, crystalloid deposits upon the boiler plates. The plates, being no longer in contact with water, are over-heated, and destroyed.

Where the quantity of these salts is not very large (12 grains per gallon, say) boiler doctors will be found very effective. They either form with the salts other salts soluble in hot water; or precipitate them in the form of soft mud, which does not adhere to the plates, and can be sludged out from time to time. The selection of a doctor must depend upon the composition of the water, and to be thoroughly satisfactory it should be introduced regularly with the feed, not once for all at each periodic cleaning of the boiler.

Examples :—

The deposition of carbonate of lime can be prevented by dissolving sal-ammoniac in the water (it will, however, damage the plates). The chloride of calcium and carbonate of ammonia produced being soluble in water :—

$$\text{Carbonate of lime} + \text{Sal-ammoniac} = \text{Chloride of calcium} + \text{Carbonate of ammonia.}$$

$$(71.)\ CaCO_3 + 2NH_4Cl = CaCl_2 + (NH_4)_2CO_3.$$

Sulphate of lime by carbonate of soda. The sulphate of

soda produced is soluble in water; and the carbonate of lime falls down in grains, does not adhere to the plates, and may, therefore, be blown out or gathered into sediment collectors :—

$$\text{(72.)} \quad \underset{\text{of lime}}{\text{Sulphate}} + \underset{\text{of soda}}{\text{Carbonate}} = \underset{\text{soda}}{\text{Sulphate of}} + \underset{\text{lime.}}{\text{Carbonate of}}$$
$$\text{CaSO}_4 + \text{Na}_2\text{CO}_3 = \text{Na}_2\text{SO}_4 + \text{CaCO}_3$$

Sodium phosphate will decompose the sulphates of lime and magnesia :—

$$\text{(73.)} \quad \underset{\text{of lime}}{\text{Sulphate}} + \underset{\text{phosphate}}{\text{Sodium}} = \underset{\text{phosphate}}{\text{Calcium}} + \underset{\text{soda}}{\text{Sulphate of}}$$
$$\text{CaSO}_4 + \text{Na}_2\text{HPO}_4 = \text{CaHPO}_4 + \text{Na}_2\text{SO}_4$$

$$\text{(74.)} \quad \underset{\text{of magnesia}}{\text{Sulphate}} + \underset{\text{phosphate}}{\text{Sodium}} = \underset{\text{of magnesia}}{\text{Phosphate}} + \underset{\text{soda}}{\text{Sulphate of}}$$
$$\text{MgSO}_4 + \text{Na}_2\text{HPO}_4 = \text{MgHPO}_4 + \text{Na}_2\text{SO}_4$$

Where the quantity of salts is large, boiler doctors are not of much use. Some other source of supply must be sought, or the bad water purified before it is allowed to enter the boilers. And as the damage done, especially to locomotive boilers, by unsuitable water is enormous, it is worth while going to considerable expense to obtain a good supply.

Pure water may be obtained by collecting rain, or condensing steam by means of surface condensers. The water thus obtained should be mixed with a little bad water, as, undiluted, pure water corrodes iron; or, after each periodic cleaning, the bad may be used for a day or two to put a skin upon the plates.

The carbonate of lime and magnesia may be precipitated either by heating the water or by mixing milk of lime (Porter-Clark process) with it, the water being then filtered. And Maxwell-Lyte and Maignen have each invented a process for dealing with the sulphates upon a large scale.

4th. *Corrosion* may be produced by the use of pure water, or by the presence of acids in the water, and, perhaps, in the engine cylinder, by the action of high pressure steam upon the grease, resulting in the production of fatty acids.

Horse-Power of Boilers.

The remedy for the first has been pointed out above. Acid water may be neutralised by the addition of lime.

Horse-Power of Boilers.

1st. *Nominal horse-power* relates more to the size than to the power of the boiler.

Experience shows that 1 square yard of effective heating surface, and from 0·75 to 1·0 square foot (according to quality of coal), of fire-grate are required for each cubic foot of water at 60° evaporated per hour into steam of any pressure, and this is considered to be equal to one NHP. In a flue boiler this involves the consumption of about 7 to 9 lbs. of good steam coal, or about 12 lbs. of rough small. The consumption of fuel depends principally upon the draught:—

A Cornish boiler, with slow combustion and very sluggish draught, consumes about 5 lbs. of good steam coal per square foot of grate per hour.

A Cornish or Lancashire boiler, with good chimney draught, about 14 lbs.

A cylindrical boiler, with good chimney draught, about 20 lbs.

A locomotive boiler, and others with strong steam blast, 100 lbs.

If NHP = Nominal horse-power.
  A = Area of effective heating surface in square yards.
  F = Fire-grate area in square feet.

(75.) $NHP = \sqrt{AF}$.

In a boiler fired externally, A, the effective heating surface, is about $\frac{3}{8}$ of the whole surface of the boiler. In the case of flue boilers, add to the above $\frac{1}{2}$ of the surface of the flues.

If L = Length of a boiler or any cylindrical vessel in feet.
  D = Diameter do. do.
  S = Surface of the sides in square feet.

(76.) $S = 3\cdot1416 DL$

2nd. *Actual horse-power*, i.e., the ability of a boiler to supply an engine working with a given indicated horse-power, is not so easily estimated, the required size depending upon the kind of engine, the pressure of steam, quality of the coal, skill of the fireman, &c. Roughly speaking, the boilers about our collieries, viz., egg or flue boilers supplying non-condensing engines, working with but little expansion, and at pressures not more than 40 lbs., will give an actual horse-power $1\frac{1}{2}$ times their nominal horse-power. Where high pressure, condensation, or much expansion are used, boilers may be worked to four times their NHP.

If the consumption and pressure of the steam are known, as in the case of designing a boiler to supply any given engine, find the cubic feet of water that must be evaporated per hour from Table XVI. This will be the NHP of the boiler required, and the size, therefore, can be got from formula (75).

This table also shows the amount the steam will fall in temperature from the beginning to the end of the stroke when used expansively.

TABLE XVI.

SHOWING PRESSURE, TEMPERATURE, WEIGHT, VOLUME, TOTAL HEAT, AND LATENT HEAT OF SATURATED STEAM.

Total Pressure. Lbs.	Temperature, Fah.	Weight in Ozs. per Cubic Foot.	Volume compared with Volume of Water that has produced it.	Total Units of Heat per lb. from 32°.	Latent Heat per lb.
10	194	0·4208	2,375	1,141	979
11	198	0·4608	2,167	1,143	976
12	202	0·5008	1,994	1,144	973
13	206	0·5408	1,846	1,145	971
14	210	0·5808	1,720	1,146	968
15	213	0·6208	1,609	1,147	965
16	217	0·6608	1,512	1,148	963
17	220	0·7008	1,427	1,149	961
18	223	0·7408	1,350	1,150	959
19	226	0·7792	1,282	1,151	958
20	228	0·8192	1,220	1,152	956
21	231	0·8576	1,165	1,153	953
22	234	0·8976	1,113	1,153·5	951
23	236	0·9376	1,067	1,154	950
24	238	0·9760	1,024	1,155	948
25	240	1·0144	985	1,156	946
26	243	1·0544	948	1,156·5	945

TABLE XVI.—continued.

Total Pressure. Lbs.	Temperature, Fah.	Weight in Ozs. per Cubic Foot.	Volume compared with Volume of Water that has produced it.	Total Units of Heat per lb. from 32°.	Latent Heat per lb.
27	245	1·0928	915	1,157	943
28	247	1·1312	883	1,157·5	942
29	249	1·1696	854	1,158	940
30	251	1·2080	827	1,159	939
31	253	1·248	801	1,159·5	938
32	254	1·286	767	1,160	936
33	256	1·325	755	1,160·5	935
34	258	1·360	734	1,161	934
35	260	1·400	714	1,161·5	933
36	261	1·438	695	1,162	931
37	263	1·477	677	1,162·5	930
38	264	1·515	660	1,163	929
39	266	1·553	644	1,163·5	928
40	268	1·590	628	1,164	927
41	269	1·628	614	1,164·5	926
42	271	1·665	600	1,165	925
43	·272	1·704	587	1,165·5	924
44	273	1·741	574	1,166	923
45	275	1·779	562	1,166·5	922
46	276	1·816	551	1,166·8	921
47	277	1·853	539	1,167	920
48	279	1·891	529	1,167·5	919
49	280	1·928	519	1,167·8	918
50	281	1·965	509	1,168	917
51	283	2·001	499	1,168·5	916
52	284	2·038	490	1,168·8	915
53	285	2·075	482	1,169	914
54	286	2·112	473	1,169·5	913
55	287	2·149	465	1,169·8	912
56	288	2·185	457	1,170	911·5
57	290	2·222	450	1,170·5	911
58	291	2·259	443	1,170·8	910·5
59	292	2·294	436	1,171	910
60	293	2·331	429	1,171·;	909·5
65	298	2·512	398	1,173	906
70	303	2·689	372	1,174 8	901
75	308	2·867	349	1,176·5	898
80	312	3·043	329	1,177·8	895
85	316	3·217	311	1,179·2	891
90	320	3·390	295	1,180·5	888
95	324	3·560	281	1,181·5	885·7
100	328	3·728	268	1,182·5	883·7

Surface condensers.—For ordinary colliery engines (*i.e.*, for engines working at pressures of about 40 lbs. and with very little expansion), for each indicated horse-power, 4 square feet of tube surface are required, and $2\frac{1}{2}$ gallons of cooling water per minute.

Injector condensers.—To the quantity of water theoretically required about 30 per cent. should be added.

Let Q = lbs. of condensing water theoretically required per lb. of steam to be condensed.
  H = Total heat of exhaust steam (see Table XVI.).
  T = Temperature of water of condensation.
  t = Temperature of condensing water.

Then:—

$$(77.) \quad Q = \frac{H - T}{T - t}$$

The cost of repairs per annum is (roughly), for:—
 An egg-ended boiler, £13.
 A Cornish boiler, £17.
 A Lancashire boiler, £20.

If the feed-water be bad the cost will be much higher.

Chimneys:—

Let H = Height in feet.
  L = Length of flue and height of chimney in feet.
  V = Velocity with which the gases travel in the chimney in feet per second.
  D = Inside dia., if round, or length of side, if square, in feet.
  h = Head in feet of air, of the temperature of the air inside the chimney, required to produce the draught.
  T = Absolute temperature of gases discharged by chimney.
  t = Absolute temperature of air before entering the furnace.

Then:—

$$(78.) \quad h = \frac{V^2}{64}\left(13 + \frac{\cdot 048 L}{D}\right)$$

$$(79.)\ H = \frac{h}{\left(0\cdot 96\,\dfrac{T}{t} - 1\right)}$$

In practice 300 cubic feet of air will be required per lb. of coal burned; and the absolute temperature and volume of the discharged gases will be about double the absolute temperature and volume of the air before entering the furnace. About 16 feet per second is a fair value for V.

Authorities.—"A Practical Treatise on Heat," Box; "The Mechanical Engineering of Collieries," Percy; "Pocket-Book of Engineering Formulæ," Molesworth; "Steam Boilers," Armstrong; "The Mines Act, General and Special Rules; The Steam Engine," Rankin; "The Theory of the Steam Engine," Baker; "Steam and the Steam Engine," Clark. Trans. N.E.I., xvii. and xxxii.; "Tall Chimney Construction," Bancroft; "The Workshop Companion," Templeton; "The Steam Engine," Cotterill.

TRANSMISSION OF POWER.

It is impossible to place boilers in a mine inbye at long distances from the shaft. The economical transmission of power, therefore, to long distances is a matter of great importance.

Wooden spears may be used for distances of 300 or 400 yards, where the road is straight; but for distances greater than this our choice is confined to *compressed air, wire ropes, steam*, and, in certain cases, *water*. Some day, possibly. *electricity* may be used for this purpose.

Compressed air.

Theory.—We note in our practical experience of compressors and air engines that :—

1. If you compress air (*i.e.*, do work upon it), you will raise its temperature, and the rise in temperature will be an exact measure of the work done upon the air.

2. If you expand air against any opposing force (*i.e.*, get work out of it), you will lower its temperature, and the fall in temperature will be an exact measure of the work got out of the air.

3. If you raise the temperature of air you will increase its expansive force.

4. If you lower the temperature of air you will decrease its expansive force.

These phenomena can be easily explained if we assume the truth of the dynamical theory of gases. It is supposed that the particles of air are flying about in all directions; and that, if they were not retained by any force, they would fly apart into infinite space. The particles strike against one another, and against the sides of the vessel that contain them; and, being perfectly elastic, they rebound with a velocity equal to the velocity of collision. The energy of the particles is the heat that the air possesses. To increase the temperature is to increase the energy, *i.e.*, to increase the velocity of movement of the particles. To decrease the temperature is to decrease their velocity. In other words, then, "Heat is a mode of motion."

Suppose we have a cylinder full of air and the piston be pushed down. The particles of air striking the piston will rebound from it with their original velocity, plus an increased velocity due to the velocity of the piston. That is to say the advancing piston striking the particles will increase their velocity, which we have just seen is equivalent to an increase of temperature. Conversely if the piston be pushed back again by the expansive force of the air, each particle that strikes the piston gives up a portion of its energy to it, and rebounds with a decreased velocity, *i.e.*, there will be a decrease in the temperature of the air.

Changes in the temperature, pressure, and volume of air are governed by the following laws :—

Let P_1 V_1 and T_1 = the initial pressure, volume and absolute temperature of a given weight of air.

P_2, V_2, and T_2 = the final do. do.

Then—

(80.) At constant temperature $P_1V_1 = P_2V_2$.

(81.) At constant pressure $\dfrac{V_1}{V_2} = \dfrac{T_1}{T_2}$.

(82.) At constant volume $\dfrac{P_1}{P_2} = \dfrac{T_1}{T_2}$.

If air be expanded or compressed adiabatically, the following relations hold good :—

(83.) $\dfrac{P_2}{P_1} = \left(\dfrac{V_1}{V_2}\right)^{1\cdot 408}$

(84.) $\dfrac{T_2}{T_1} = \left(\dfrac{V_1}{V_2}\right)^{\cdot 408} = \left(\dfrac{P_2}{P_1}\right)^{\cdot 29}$

The units of work = U, required to compress a volume of air = V_1, to a volume of air = V_2; or to compress a volume = V_1, from P_1 to P_2.

1st, *isothermally* : *i.e.*, at constant temperature, are :—

(85.) $U = P_1 V_1 \text{ hyp log.} \dfrac{V_1}{V_2}$.

2nd, *adiabatically* : *i.e.*, without loss of the heat due to compression, are :—

(86.) $U = \dfrac{P_1 V_1{}^{1\cdot 408}}{\cdot 408} \left\{ \dfrac{1}{V_2{}^{\cdot 408}} - \dfrac{1}{V_1{}^{\cdot 408}} \right\}$.

As, when air is compressed adiabatically, the rise in its temperature is an exact measure of the work done upon it; the units of work required to compress it can be calculated from the rise in temperature. This increase of temperature $T_2 - T_1$ is given by (84); and the units of work = this quantity × weight of the air in lbs. × specific heat of air at constant volume expressed in foot lbs., viz. :—130·3.

Thus if W = weight of the air in lbs. :—

(87.) $U = (T_2 - T_1)\ 130\cdot 3\ W$.

Conversely in order to calculate the units of work given out by compressed air when expanded isothermally or adiabatically the same formulæ (*i.e.*, 85 to 87) should be used.

Practice.—In order to compress air a Ram Compressor may be used if there be a plentiful fall of water ("Power of Water," Weale's series, No. 82); or the compressors of Sommeiller, Colladon, &c., described by André in his "Mining Machinery," where a fall of water is not available.

Loss due to *physical properties of air.*—Cooling of compressor with cold water, heating of air-engine with hot water or with steam ; clearance spaces.

Loss due to *friction in conducting pipes.*—The formula for calculating this is of the same form for all fluids, and is given on p. 62, formulæ (97) to (99). P—p. is the loss of pressure between the compressing engine and the air engine, and d is the weight in oz. of a cubic foot of the compressed air, which last may be got from formula (144), bearing in mind that two inches of the barometer are equal to 1 lb. of pressure.

The greater the pressure of the air the less its efficiency.

Mr. W. Daniell found that, with a pressure of 19 lbs. above the atmosphere, the compressed air gave 45·8 per cent. of the horse-power of the compressing steam-engine; with a pressure of 40 lbs. above the atmosphere, 25·8 per cent. only.

The following books may be consulted :—" Spon's Dict. of Eng. Supp.;" Trans. N.E.I., xxi., xxii., xxxi. ; " Mining Machinery," André; " Power of Water," Glynn; " Transmission of Power by Compressed Air," Zahner.

Wire Ropes.

Wire rope, or Telodynamic, transmission, is not used in England so much as we think it might be; though the confined passages of a mine will not admit of the large sheaves necessary for a perfect installation.

HP = Horse power transmitted to driven sheave.
S = Speed of rope in feet per minute.
P = Force, or pull of rope, in lbs.
K = Coefficient of efficiency depending upon the dimensions of the sheaves, the distance, &c.

(88.) $$HP = \frac{KPS}{33,000}.$$

Where the main and tail rope system of haulage is in use, the return sheave may be utilised for working a pump or other machinery. It should be cleaded with wood, which can be renewed from time to time. If, however, a special installation has to be made, it will be better to use an endless rope. The driving and driven sheaves should be of large diameter, 200 times that of the rope is found to be the best proportion; but little more than half this can be attained in mines. They should be cleaded with wood —willow is the best—gutta-percha, or leather set on edge. The intermediate sheaves may be a foot in diameter, though the larger they can be made the better, up to six feet; the rope of tough, flexible steel, made of a large number of small wires. With large main sheaves (12 to 15 feet), its speed should be from 30 to 40 miles an hour, and it is under these conditions, viz. :—A rope running at a great speed, under a small strain—that a rope transmission is most effective. With the small sheaves (say main sheaves six feet, and intermediate sheaves 12 inches), that can only be used in mines, about 12 miles an hour would be a suitable speed; and we know of one transmission, running at only 6 miles an hour, that was fairly successful. The objection to these slow speeds is the great strain upon the rope. Binding sheaves are not required; but it is convenient to have the driving or the driven sheave set upon a sliding carriage to compensate for the stretching of the rope.

The loss of power due to friction, &c., in a carefully-proportioned, above-ground transmission, appears to be only about $2\frac{1}{2}$ per cent., and $\frac{3}{4}$ per cent. in addition for each 1,000 yards. In the mine, however, such perfection cannot be attained on account of the small diameter of the sheaves and rollers, and the numerous curves. For distances of from 2,000 to 3,000 yards, 30 to 40 per cent. of the horse-power of the engine will be absorbed in driving the ropes.

Authorities:—Proc. I.M.E., 1874, p. 56; Trans. N.E.I., xvii.; "Transmission of Power by Wire Ropes," Stahl; "The Engineer," xxiii. and xxxvii.

Steam.

The losses in a steam transmission are of two kinds, viz.:—

1. Loss of steam from condensation.
2. Loss of pressure from friction.

The loss of heat, and consequent condensation, is due to two causes—radiation and contact with air.

Loss by Radiation.

Let U = units of heat lost by radiation, per hour.
S = surface of covered pipe in square feet.
D = difference of temperature in degs. Fahrenheit between surface of covered pipe and drift or shaft sides.
R = (see Table XVII.)

Then—
(89.) $U = 0{\cdot}74 DSR$.

Let L = lbs. of steam condensed per hour.

H = latent heat of steam (see Table XVI.) at the mean pressure in the pipe.

Then—

(90.) $L = \dfrac{U}{H}$.

TABLE XVII.

The Ratio of Heat Emitted or Absorbed at Different Temperatures.

Let R = ratio of loss of heat.

t = cent. temperature drift sides.

T = cent. difference of temperature between pipe surface and drift sides.

Then—

(91.) $R = \dfrac{124{\cdot}72 \times 1{\cdot}0077^{t} \times (1{\cdot}0077^{T} - 1)}{T}$.

Reduced to Fahrenheit's scale this formula gives the following, Table XVII. :—

T.	*t.*							
	59°	63½°	68°	72½°	80°	86°	91°	95°
Degs.	R.	R.	R.	R.	R.	R.	R.	R.
18	1·120	1·14	1·165	1·178	1·223	1·254	1·271	1·300
27	1·14	1·159	1·185	1·190	1·247	1·276	1·305	1·327
36	1·16	1·177	1·206	1·202	1·272	1·299	1·330	1·352
45	1·18	1·20	1·229	1·231	1·297	1·323	1·357	1·381
54	1·20	1·23	1·251	1·261	1·323	1·348	1·385	1·406
63	1·225	1·25	1·276	1·289	1·350	1·375	1·407	1·435
72	1·25	1·283	1·302	1·317	1·377	1·403	1·429	1·464

Loss by Contact with Air.

This will be greater with horizontal than with vertical pipes.

Horizontal Pipes.

Let U_1 = units of heat lost per hour by a horizontal pipe from contact with air.

S = surface of covered pipe in square feet.

R_1 = (see Table XVIII.)

D_1 = difference of temperature in degrees Fahrenheit between surface of covered pipe and the air.

A = (see Table XIX.)

Then—

(92.) $U_1 = D_1\ R_1\ A\ S$,

And the lbs. of steam condensed can be found from (90).

TABLE XVIII.

THE RATIO OF HEAT EMITTED OR ABSORBED BY CONTACT WITH AIR WITH GIVEN DIFFERENCES OF TEMPERATURE.

Let R_1 = ratio of loss of heat.

t = difference of temperature of the pipe surface and the air in degrees cent.

Then—

(93.) $R_1 = \dfrac{0\cdot 552 \times t^{1\cdot 233}}{t}$.

Reduced to Fahrenheit's scale, this formula gives the following, Table XVIII. of values of R_1:—

t.	R_1.	t.	R_1.
Degrees.		Degrees.	
9	0·782	54	1·219
18	0·943	63	1·263
27	1·037	72	1·305
36	1·109	81	1·341
45	1·168	90	1·372

TABLE XIX.

If r = radius of horizontal covered pipe in inches,

(94.) $A = 0\cdot 421 + \dfrac{0\cdot 307}{r}$.

LOSS BY CONDENSATION.

r	A.	r	A.	r	A.	r	A.
2	0·5745	3¼	0·5154	4½	0·4892	6½	0·4682
2¼	0·5574	3½	0·5087	4¾	0·4856	7	0·4648
2½	0·5440	3¾	0·5028	5	0·4824	7½	0·4619
2¾	0·5326	4	0·4978	5½	0·4768	8	0·4593
3	0·5230	4¼	0·4930	6	0·4722	9	0·4551

Vertical Pipes.

Let U_{11} = units of heat lost per hour by a vertical pipe from contact with air.

A_1 = (see Table XX.)

Then— S, R_1, and D_1, as in (92).

(95.) $U_{11} = D_1 R_1 A_1 S$,

And the lbs. of steam condensed can be found from (90.)

TABLE XX.

If r = radius of vertical covered pipe in inches.

h = height of vertical pipe in inches.

Then—

(96.) $A_1 = \left\{ \cdot 726 + \dfrac{\cdot 2163}{\sqrt{r}} \right) \times \left(2 \cdot 43 + \dfrac{5 \cdot 49}{\sqrt{h}} \right\} \times \cdot 2044.$

This formula gives the following Table XX., taking A_1 in feet.

Radius in Inches.	Height of Pipe in Feet.			
	50	100	200	300
r	A_1.	A_1.	A_1.	A_1.
2	0·4769	0·4650	0·4571	0·4534
2½	0·4676	0·4560	0·4478	0·4442
3	0·4614	0·4500	0·4419	0·4384
3½	0·4562	0·4448	0·4368	0·4333
4	0·4526	0·4412	0·4333	0·4299
4½	0·4491	0·4378	0·4300	0·4266
5	0·4462	0·4352	0·4273	0·4239
5½	0·4437	0·4328	0·4250	0·4216
6	0·4416	0·4298	0·4220	0·4186
6½	0·4398	0·4289	0·4212	0·4178
7	0·4380	0·4272	0·4196	0·4162
7½	0·4366	0·4257	0·4181	0·4147
8	0·4352	0·4244	0·4168	0·4134
9	0·4330	0·4220	0·4146	0·4112

By means of the above formulæ and tables, the quantity of steam that will be condensed (that is to say, the quantity of steam that must be produced by the boiler in addition to that required to drive the engine,) in the range of pipes can be easily calculated if only the surface temperature of the pipes, of the air, and drift sides be known. How these may be obtained will be presently pointed out.

Loss by Friction.

The laws governing the resistance that fluids meet with in passing through iron pipes (and other conduits also; in which, however, we are not now interested) do not appear to be thoroughly understood. M. Stockalper, however, found, from experiments upon the flow of compressed air through pipes made at the Mont Cenis tunnel, and published in the "Revue Universelle des Mines," Sér. 2, Vol. VII., p. 257, that Darcy's formula for the flow of water through iron pipes, reduced in the ratio of the density of air to that of water, gave satisfactory results. Acting upon these suggestions, the author has made use of Darcy's formula for the flow of water, after having converted it into British units, as follows:—

Let P = boiler pressure in lbs. per square inch.

p = pressure required at engine in lbs. per square inch.

l = length of pipe in yards.

d = weight of 1 cubic foot of the fluid in oz. (for steam, see Table XVI.)

D = diameter of pipe in inches.

Q = cubic feet of the fluid passing per second. (See below.)

a = (See Table XXI.)

Then—

$P - p$ = loss of pressure between boiler and engine;

And—

(97.) $P - p = \dfrac{l a Q^2 d}{1,000,000}.$

(98.) $Q = \sqrt{\dfrac{1,000,000 \, (P - p)}{l a d}}.$

(99.) $a = \dfrac{1,000,000 \, (P - p)}{l Q^2 d}.$

There is some difficulty in finding the value of this quantity Q, the mean volume; but assuming that, in a pipe of uniform section, with no very great variation in temperature, the condensation takes place uniformly from end to end; and that the loss from leakage is inappreciable:—

If V = volume of steam in cubic feet per second produced by the boiler.
v = volume of steam in cubic feet per second consumed by the engine.
Q = mean volume in cubic feet per second. passing through the pipe.

Then, assuming that V—v is the volume of steam lost by condensation; that is to say, neglecting leakage:—

$$(100.)\ Q^2 = \frac{V^2 + Vv + v^2}{3},$$

and formula (97) becomes—

$$(101.)\ P - p = \frac{l\,a\,d}{1,000,000} \times \frac{V^2 + Vv + v^2}{3}.$$

TABLE XXI.

VALUES OF a FOR DIFFERENT INTERNAL DIAMETERS D, OF PIPES IN INCHES.

$$(102.)\ a = \frac{306{,}703{,}494 b}{D^5};\ \text{and}$$

$$(103.)\ b = \cdot 000507 + \frac{\cdot 00050946}{D}$$

Internal Diameter of Pipe in Inches.	a.	Internal Diameter of Pipe in Inches.	a.
1¼	91,960	5½	36·54
2	7,302	6	23·29
2½	2,232	6½	15·47
3	812	7	10·58
3½	381	8	5·34
4	190	9	2·927
4½	103·7	10	1·717
5	59·7	11	0·9872

The Design of a Steam Transmission.

The only difficulty that can arise in making use of these formulæ for the purpose of determining the size of pipes and boiler power required for any proposed transmission, lies in the estimation of the surface temperature of the covered pipe. This will depend upon the composition used, and its thickness; and upon the temperatures of the steam, the air, and drift sides. As it is independent of the diameter and length of the pipe, the simplest plan is to make an experiment by covering three or four yards of pipe with the composition to be used. Or reference may be made to a very valuable series of experiments upon various non-conducting compositions, carried out by Mr. Bird, Assoc. Sc., and read before the N. of England Institute. See Vols. XXIX., XXXI., and XXXII.

The following, Table XXII., shows the results of some experiments made by the author with Wormald's composition. It will be noted that the differences of temperature do not vary much :—

TABLE XXII.

EXPERIMENTS WITH WORMALD'S COMPOSITION.

Thickness of Composition.	Approximate Temperature of the Steam.	Temperature of Surface of Covered Pipe.	Temperature of the Air.	Difference of Columns 3 and 4.	Temperature of Drift Sides.
Inches.	Degrees.	Degrees.	Degrees.	Degrees.	Degrees.
$1\frac{1}{8}$	281	122	77	45	72
$1\frac{1}{4}$	275	101	62	39	62
$1\frac{3}{8}$	275	100	62	38	62
$1\frac{7}{16}$	281	121	77	44	72
$1\frac{7}{16}$	271	120	77	43	72
$1\frac{11}{16}$	285	120	79	41	74
$1\frac{3}{4}$	287	$132\frac{1}{2}$	91	$41\frac{1}{2}$	89
$1\frac{7}{8}$	287	132	94	38	$93\frac{1}{2}$
$1\frac{7}{8}$	287	109	77	32	open air
$1\frac{7}{8}$	286	107	76	31	do.
$2\frac{1}{8}$	287	107	77	30	do.
$2\frac{1}{8}$	286	104	76	28	do.

Having determined the temperatures in one or other of these ways, the steam required to supply the condensation can be readily obtained from formulæ (89) to (95). Experience seems to show that a little under 10 per cent. must be added to this for condensation at the steam traps and expansion joints.

The volume required for the engine is of course known, and this (the engine volume) added to the condensation volume gives the gross quantity to be supplied by the boiler, and consequently the boiler power required.

The mean volume squared passing through the pipe is got from the engine volume and condensation volume by formula (100), and finally the size of the pipes from formula (99) and Table XXI.

Practical Details.

Provision must be made for carrying off the water of condensation, and for expansion of the steam pipes. The first is well understood, and I would only suggest that a trap be placed as near the boiler as possible, to intercept the water carried over by priming. It was found at Broomhill that whereas the trap next the boiler gave a gallon per 7·3 yards of pipe, the second trap from the boiler gave a gallon per 15·8 yards.

For the low pressures usually adopted at collieries (say not more than 45 lbs. above the atmosphere) the ordinary stuffing-box expansion joint answers admirably; but with higher pressures there is considerable difficulty. The New York Steam Company (pressure 80 lbs.) have made a great many experiments upon expansion joints, and finally settled upon a modification of the diaphragm joint. It is made of discs of copper 0·04 inches thick, corrugated concentrically, and supported on radial backing plates, which prevent the diaphragm from being distended to rupture by the pressure.

Provision must be made for dealing with the exhaust steam. If the engine is used for pumping, and there be sufficient water, the simplest plan is to turn the exhaust direct into the suction pipe. By this means not only is the steam killed but a vacuum is obtained, and the engine made more efficient. At East Howle Colliery, instead of turning the exhaust direct into the suction, they carry the exhaust pipe some thirty yards inside the rising main, and then turn it into the suction pipe. By this means they consider that they get a more perfect condensation than if the exhaust steam were turned direct into the suction; and they certainly pass cold water through the pump instead of hot, which is an undoubted advantage.

Authorities :—Box on "Heat;" Trans. N.E.I., Vols. XXIX., XXXI., XXXII., XXXV., and XXXVI.

Summary.

A *compressed air* installation requires a large capital expenditure: but once established, it is not expensive to maintain where fuel is cheap. It is the most handy form of transmission for mining purposes, as the power can be readily split up by means of branch pipes, and carried in small quantities to numerous points. In addition, the exhaust air improves the ventilation, cools the mine, and can be used for clearing away gas.

A *wire rope* transmission is much less costly in the first case than compressed air; and, if properly laid out, a large quantity of power may be led to one or two points with little loss of useful effect. But the cost of maintenance is larger and the power cannot be readily carried in small quantities to many points.

Steam is not to be recommended, except in special cases; for, however carefully the pipes may be covered and the exhaust dealt with, there will always be a considerable escape of heat, which is very inconvenient in the confined passages of a mine.

Electricity is not likely, we think, ever to compete successfully with wire ropes or with air for the transmission of power in mines; where the maximum distance is only about three miles. It will, however, no doubt be applied some day for the transmission of power to very great distances. It is immaterial whether the road be straight or crooked; whether the work to be done be concentrated, or distributed in small quantities over many points.

MAGNETISM AND ELECTRICITY.

Practical Electro-Magnetic Units.

Electromotive-force (and Potential) :—The Volt = 10^8 absolute units ; and is from 5 to 10 per cent. less than the E.M.F. of one of Daniell's cell.

Resistance :—The Ohm = 10^9 absolute units ; and is about equal in resistance to 48·5 metres of pure copper wire, 1 mm. dia., at 0° cent.

Current :—The Ampère = 10^{-1} absolute units ; and is that furnished by an E.M.F. of one volt, through a resistance of one ohm.

Quantity :—The Coulomb = 10^{-1} absolute units; and is the quantity of electricity passing per sec. across any section of a circuit through which a current of one ampère is flowing, *i.e.*, one Ampère = one Coulomb per second.

Capacity :—The Farad = 10^{-9} absolute units ; and is the capacity of a conductor which a charge of one coulomb raises to a potential of one volt.

To calculate the Horse-power of a current :—

Let HP = Horse-power.
A = Ampères.
V = Volts.

Then :

(104.) $$HP = \frac{AV}{746}.$$

One Ampère volt = one Watt.
∴ one Horse-power = 746 Watts.

Compass Surveying.

It must be remembered that the needle does not point to true north ; but, in Great Britain, at the present time, to the west of true north : and that the angle contained by these two straight lines; viz., the true north and south line, and the magnetic north and south line, differs at different places. This angle, at any place, is called the declination—or, more commonly, the variation—of the needle for that place. The variation is the same along a straight line drawn through the North of England coal-field, skirting the W. side of Durham, and the E. side of Newcastle. At the present time, the magnetic variation on this line is $19° 29'$ W. of N. At collieries in this coal-field E. and W. of this line, the *maximum* difference is $15'$; viz., $15'$ less on the E., $15'$ more on the W. The variation is decreasing at the rate of about $7'$ a year, so that, in 1890, the variation along the above line will be $19° 29' - (7' \times 3) = 19° 29' - 0° 21' = 19° 8'$ W. of N.

TABLE XXIII.

PLAN EQUIVALENTS.

Inches per Inch, *i.e.*, Scale.	Feet per Inch.	Yards per Inch.	Chains per Inch.	Miles per Inch.	Acres per Square Inch.	Inches per Mile.
792	66·0	22·0	1·0	0·0125	0·10	80·0
1,584	132·0	44·0	2·0	0·0250	0·40	40·0
2,376	198·0	66·0	3·0	0·0375	0·90	26·66
2,500	208·33	69·44	3·15	0·039	0·996	25·344
3,168	264·0	88·0	4·0	0·05	1·60	20·0
3,960	330·0	110·0	5·0	0·0625	2·50	16·00
7,920	660·0	220·0	10·0	0·125	10·0	8·0
10,560	890·0	296·66	13·33	0·166	17·77	6·0
63,360	5,280	1,760	80·0	1·0	640·0	1·0
126,720	10,560	3,520	160·0	2·0	2,560	0·5
190,080	16,840	5,280	240·0	3·0	5,760	0·333
253,440	21,120	7,040	320·0	4·0	10,240	0·25

Firing Shots

The advantages of Electric Shot Firing are:—Shots fired simultaneously thereby more effective. Saving of time. Safer, because not fired until all men are out of the way; and should a shot miss, it cannot fire afterwards in the face of the workman examining it.

Two kinds of fuse are used, viz. :—
(1.) *Tension fuse* fired by a frictional, or magnetic-electric machine.
(2.) *Quantity fuse* fired by a voltaic battery.

> (See " Practical Treatise on Coal Mining," by André, p. 208.)

Signalling.

The Single-wire System, in which signals can only be sent from certain fixed stations.

The Double-wire System, in which signals can be sent from any point, by making contact between the two wires.

In both these systems, the electric current is made to ring a bell; and a convenient form of battery is the Leclanché, as it does not require much attention, and is not liable to speedy exhaustion.

Lighting.

The Incandescent Lamps of Maxim, Swan, &c., are used, both for lighting at bank and below ground. These are stationary lights, and the current is produced by a dynamo-electric machine, driven by any convenient water, or steam-engine, with which dynamo the lamps are connected by means of wires. No moveable lamp has yet been used in mines, as there is a difficulty about the battery. Mr. Swan, however, is engaged upon this problem and has invented a portable miners' electric lamp; but, as at present constructed, it is too costly for practical use. One great advantage of the incandescent lamps, is that, burning only in a vacuum, they cannot (if reasonable precautions are taken to prevent breakage of wires) fire gas.

In an installation of 50 lamps and upwards, each lamp of 20 candles, one horse-power is required per 8 to 10 lamps.

Swan's 20-candle lamps, *joined in series*, require an E.M.F. of 45 to 60 volts per lamp, with a current of one ampère. *Joined parallel,* they require a current of one ampère per lamp, with an E.M.F. of from 45 to 60 volts.

Geissler Tubes, as in the lamp of Benoit and Dumas, have been suggested; but the lamp is heavy, and the light small.

Transmission of Power.

A little has been done in France in this direction. The efficiency of electricity compared favourably with the alternative, compressed air. (See Trans. N.E.I., xxxi., Abs., pp. 9—11, xxxii. Abs., pp. 13, 14, xxxiii. Abs., p. 71, and xxxiv.)

Fire-Damp Detectors.

Liveing, Ansell, Maurice, Swan, and Somzée have each contrived an electric fire-damp detector. That of Liveing seems the most practical, and is in regular use at Pagebank and other collieries. They are all described in the Trans. N.E.I.

Danger.

From shocks, there is none when a continuous current of not more than 200 volts, or an alternating current of not more than 75 is used. This is, I believe, the limit adopted by the Board of Trade, and errs slightly on the right side.

From sparks firing gas, should any be present, must always be guarded against. This may occur at the commutator; or from a broken wire, the spark leaping across the space between the two ends at the moment of rupture.

The following books may be consulted:—" Electricity and Magnetism," Silvanus Thompson; "Electricity," John T. Sprague; Trans. N.E.I., Vols. xxx., xxxi., and xxxvi.

SEARCH FOR MINERALS.

Mineral deposits. { Irregular. Regular. { 1. Lying at high angles by costeaning. 2. Lying at low angles by boring.

1. *Costeaning* :—A simple process requiring no machinery.

2. *Boring* :—Theoretically, borings, of one or more holes, should give us the following information, viz. :—

One hole :—Vertical distance from the surface to the deposit; thickness of the deposit; quality of the deposit.

Three holes :—(In addition to the above) amount of dip; direction of dip.

To find the amount and direction of dip of a bed, by means of three bore-holes.

Let A, B, and C, be the three bore-holes.
 S = Angle of dip of bed.
 V = Angle between the strike of the bed and AB.
 a = Distance from A to B.
 a' =   Do.   A to C.
 W = Angle in a horizontal plane between AB and AC.
 d = Difference of the depths of A and B.
 d' =   Do.   A and C.

In both cases starting from the same horizontal plane.

(105.) $$\tan S = \frac{d'}{a \sin V}.$$

(106.) $$\tan V = \frac{\frac{da'}{d'} \sin W}{a - \frac{da'}{d'} \cos W}$$

Borings are made :—*By hand, a jet of water, Mather and Platt's process, the Diamond Process.*

Hand-boring :—The head-gear, rods (wood, iron, steel, ropes), and tools (cutting, clearing, extracting).

The Diamond Process.—The engine, quill, rods, sediment tube, core tube, crown, split-ring, and core.

Iron rods weigh about 21 lbs. per square inch of section per fathom; and $18\frac{1}{2}$ lbs. per square inch of section per fathom in water.

Cost of boring.—For average coal measure rocks, 7s. 6d. per fathom for the first five fathoms; 15s. per fathom for the second five fathoms; 22s. 6d. per fathom for the third five fathoms and so on, has been a standard price in the north for many years. A Scotch borer in 1883 advertised his price at 4s., in the place of 7s. 6d., as above. The price for diamond boring is 6s. per foot for the first 100 feet; 12s. per foot for the second 100 feet; 18s. for the third, and so on.

To find the cost of a bore-hole—

Let c = cost.
a = price for first step.
d = increase in price for each additional step in depth.
n = number of steps.

(107.) $c = \left\{ 2a + (n-1) d \right\} \dfrac{n}{2}.$

When n is not a multiple of a, this rule is only approximately correct.

The following deep bore-holes may be mentioned :—At Schladebach, Leipzic, 956 fathoms, 11 in. dia. at the top and 1·22 in. at the bottom; it begins in the Trias, and passes through the Permian into the Devonian formation, which point was reached in the summer of 1886. Sperenberg, Berlin, $695\frac{1}{3}$ fathoms, 12 in. dia., all in rock salt excepting the first $47\frac{1}{6}$ fathoms. Creusot, by Kind, 503 fathoms. The Rochefort bore-hole, 469 fathoms in Triassic beds. The Mondorf bore-hole, Luxemburg, 400 fathoms. The New

Saltzwerk bore-hole, Westphalia, 380 fathoms. And the Sub-Wealden bore-hole, near Battle in Sussex, 317¾ fathoms.

The following books may be consulted :—" Mining Machinery," André; " Mine Engineering," Greenwell; Trans. N.E.I., Vols. ii., x., xiii., &c. ; and " Lectures on Mining," Callon.

SINKING.

Mines Act, &c.
Prohibition of single shafts, secs. 16, 17 and 18. Fencing abandoned mines, sec. 37. Fencing and securing shafts, inspection, &c. General Rules, 18 to 38.
Pits may be sunk by means of :—
(1.) Men and machinery placed at the bottom of the pit.
(2.) Men and machinery situated at bank.

(1.) Workmen in the Pit.

The ordinary method adopted in the North of England is fully detailed in the works of André and Greenwell. The workmen, standing upon the bottom of the pit, blast out the rock, and send the excavated material to the surface by means of an engine, rope, and kibble. The sides of the shaft are retained first by temporary cribs and backing deals, and afterwards by a permanent walling. The feeders of water are drawn to the surface in the kibble or pumped by a set hanging in the shaft, and are finally tubbed back, one after the other, as they are met with. This system is all that can be desired under ordinary circumstances.

Herr Poetsch's freezing method :—See " Colliery Guardian," Nov. 16th and 23rd, 1883. May be adopted where the feeders are too excessive for the ordinary method and the ground too loose for the Kind-Chaudron process.

(2.) Workmen at Bank.

The Kind-Chaudron method :—The barrack, the engine, the rocking-lever, the spears, the free-fall, the trépan, the whimble, the tubbing, the moss box, and the concrete backing. The cost is very variable, from £74 to £338 per fathom having occurred in actual practice.

Warington-Smyth sums up an account of this method as follows :—

1. A very hazardous operation has been converted into a comparative certainty.

2. An economy of 50 to 75 per cent. has been effected on the outlay, as compared with the ordinary system carried out in the same districts. This immense gain arises mainly from :—

a. No pumps or pumping engines being required

b. The prevention, in a great degree, by the pressure of the water in the shaft, of irruptions of quicksand.

c. The employment of a comparatively small number of men, and these being, for the most part, workmen of a less highly-paid order than the regular sinkers.

d. The suppression of the vertical joints of the tubbing, whereby leakage, costs of wedging, and tendency to displacement, are avoided.

3. Risk to life is, to a great extent, eliminated by the whole of the work being done at the surface.

4. Damage to the buildings and wells of the neighbourhood is prevented by the process not requiring the drawing of water and sand. (Trans. N.E.I., xx., 198.)

The Chavette method for sinking through running sands. See Trans. N.E.I., xxxii., Abs. p. 51.

Chaudron's rule for thickness of tubbing is :—

Let E = Thickness in metres (1 m. = 39·37 inches).
R = Radius of shaft in metres.
P = Pressure in kilogrammes per square centimetre (100 metres of water weigh 10 kilogrammes per square centimetre.)

(108.) $$E = \cdot 02 + \frac{RP}{500}$$

See formula (42).

Shaft Fittings.

Walling :—This is made of bricks, fire-clay lumps, stone, or concrete. For strength see formula (42).

Tubbing :—To keep back water, of cast-iron in segments or rings.

Brattice :—Now that each colliery must have two shafts, permanent brattice is not much used. The Countess shaft, Whitehaven, is divided into four compartments by a masonry brattice; the courses are 10 inches thick and arched, so that each is self-supporting.

Guides :—These are made of wood, iron rods, iron or steel rails, or of wire ropes. Wire ropes are, perhaps, the best, on the whole, as they take up very little room in the shaft, and admit of the cage being run with great velocity.

Water rings ; Keeps ; Rappers :—Wire with lever, speaking-tube, or electric. *Shaft-gates ;* and *water rings* are also required.

The Cost of sinking and fitting up a shaft by the ordinary method depends, to some extent, upon the diameter and depth of the shaft, the rate of wages and materials; but much more upon the strata met with and the quantity of water. Some items of the cost for a 14 feet pit, 100 fathoms deep, sunk through ordinary coal measure strata and with little water, would be about:—Total labour cost of sinking and walling £25 per fathom; Contractor for sinking, including small stores £14 per fathom. Making walling beds, £6 each; Walling with fire-clay lumps, £15 per fathom walled ; Tubbing, with cast-iron segments, £90 per fathom tubbed; Plank brattice £2 10s. per fathom; Guides of wood, 15s. per fathom; Iron or steel rail guides, 50 lbs. per yard, £2 10s. per fathom; Wire rope guides, £1 5s. per fathom. And the total cost for the finished pit about £50 per fathom.

If there is a good deal of water (but still not more than can be easily mastered by the ordinary method of sinking) this price might be doubled.

Shaft Pillars, &c.

If D = Depth of shaft in fathoms.
S = Size of shaft pillars in yards.

(109.) $S = \sqrt{\dfrac{D}{50}} \times 22$ is an approximate rule. The exact size must depend upon the special circumstances of each case.

If W = Width of shaft in feet.
H = Height of hanging on place in feet.
L = Length in feet of longest prop, rail, &c., that can be taken down into the mine.

(110.) $L = (W^2 + H^2)^{\frac{1}{2}}$ $\therefore H = (L^2 - W^2)^{\frac{1}{2}}$

Some Deep Mines.

The deepest shaft in the world is I believe, that of a lead mine, the Adalbert shaft, Prizbram, 572 fathoms, 1884 (the Maria shaft, Prizbram is also about the same depth); and the deepest mine, the Viviers Reunis coal mine, at Gilly, near Charleroi, 581·5 fathoms; the depth of the shaft is 570 fathoms, but there is a staple at the bottom, 11½ fathoms deep.

The deepest mine in England is the Ashton Moss colliery, near Manchester, sunk to the Black mine, 472·5 fathoms (the total depth of the shaft is 475 fathoms). The seams dip 9″ per yard, so that parts of the workings are about 500 fathoms (1886). There are also Rosebridge colliery, Wigan, sunk to the Arley mine, 403 fathoms (the total depth of the shaft is 407½ fathoms); Dolcoath tin mine, Cornwall, 404 fathoms (1884); Harris Navigation colliery, S. Wales, 373½ fathoms to the 9-feet seam (total depth of shaft, 380 fathoms); and Dukinfield colliery, Manchester, sunk to the Black mine, 358⅓ fathoms.

The deepest mines in the North of England are Seaham colliery, 301 fathoms to the Busty seam; Silksworth colliery, 290 fathoms to the Hutton seam; and Monkwearmouth colliery, 287½ fathoms to Hutton seam.

Underground Temperature.

The temperature of the earth increases as we descend; but at what rate is not exactly known. In round numbers, the temperature at 100 feet is constant, and is equal to the mean annual temperature of the place. Below this point, the temperature increases 1° Fahr. for each 56 feet. The mean annual temperature of Newcastle is 49° Fahr. The temperature in the deeper parts of the workings at Ashton Moss is from 86° to 90°.

In some of the Comstock lode silver mines, Nevada, workings are being carried on at a depth of 336 fathoms (1877), and at a temperature up to 123°, the rock having a pretty uniform temperature of 130°. This great heat is due to the hot springs, some of which have a temperature of 158° Fahr. Chemical decomposition is thought to be, in part, at any rate, the cause of this high temperature.

Important Adits.

Gwennap, Cornwall, drains 30 square miles, is 40 miles long, including branches, and varies in depth from 30 to 90 fathoms. Ernst August, Harz, 14 miles long, including branches; greatest depth, 222 fathoms, gradient 1 in 2,000. The Blackett level, from Allendale town to the mines at Allenheads, 7 miles, of which $4\frac{5}{8}$ are completed; gradient 1 in 660.

The following books may be consulted:—"Mine Engineering," by Greenwell; "Practical Treatise on Coal Mining," by André; Trans. N.E.I., xx., xxxi.; "Lectures on Mining," by Callon; "Mines, Miners, and Mining Indus., U.S.," by Drinker; "American Jour., Sci. and Arts," 3 ser., xvii.; "Mec. Eng. of Collieries," by Percy; Report of Coal Com., 1871; Proc. I.C.E., vols. lxiv., lxxi.; and "Underground Temperatures," by Prestwich.

SYSTEM OF WORKING.

The method to be adopted depends upon a great variety of circumstances, such as the mineral worked, its mode of occurrence, &c., &c.

Minerals are found as Seams or Lodes (regular deposits). As Impregnations or Masses (irregular deposits).

Deposits vary much in thickness; confining ourselves only to those that are being, or have been worked, we find :—

Coal :—The Chapelet seam at the Hasard colliery, Liège, 1½ feet; the Three-quarter, Midgeholme, Northumberland, from 22 to 30 inches; the Maudlin, Ryhope, Durham, 12 feet; the Ten-yard seam, Dudley, upwards of 30 feet in places; the great seam, Béraudière, St. Etienne, up to 82 feet; the great seam at Bézenet up to 200 feet.

Salt :—Cheshire, 75 to 100 feet; Wieliczka, in Galicia, 100 feet; Middlesborough, up to 90 feet; Sperenberg, Berlin, 3,890 feet.

Iron-ore :—Lord Leconsfield's mine, Cleator-moor, Cumberland, 60 feet; the Lias band, Eston, 18 feet.

Silver lodes of Schemnitz, Hungary, 30 feet.

Alum Shales on the Meuse, 90 feet.

Slate Mines in the Ardennes, 60 feet.

Various as are the conditions of occurrence of mineral deposits, the systems of working them may be all classed under one or other of five heads, viz. :—

Above Ground :—

Pumping (Salt, Middlesborough).
Washing (Gold, California).
Quarrying.

Under Ground:—

Bord and Pillar.
Long Wall.
We are not interested here in the first three.

Bord and Pillar may be defined as any system of mining in which the deposit is removed in two or more workings. A portion of the deposit being left during the first working or workings in order to support the roof and sides of the excavation. (Post and Stall, Stoop and Room, Pillar and Breast are synonymous with Bord and Pillar.)

Long Wall may be defined as any system of mining in which the whole of the deposit (or, in the case of very thick deposits, a horizontal slice of it) is removed in one working; no portion being left to support the roof and sides of the excavation.

The Long Wall and Bord and Pillar methods, as adopted in the North of England for the working of seams of coal, may be shortly described and compared as follows :—

Long Wall and Bord and Pillar.

In Long Wall, a face of considerable width, say 100 to 500 yards, is opened out, and the coal is worked along the whole distance either in one lift or in steps. The roads— main-gates and cross-gates, as they are called—pass through the goaf and are supported on packs built up of the stone taken down to form height in the roads. The roof along the face is also supported on packs made from the refuse— *i.e.*, the band or folling—of the seam, and where this fails, on timber which is drawn and shifted forwards as the face advances. All superfluous stone, &c., not required for the packs, is cast back into the goaf, and one of the main elements of success in this system of working is that there should be sufficient of this to fill, more or less completely, the void left by the abstraction of the seam, so as to let down the roof evenly and gradually.

In Bord and Pillar the seam is first cut up into rectangular masses by two sets of excavations, driven at right angles to one another, and then these masses are removed in slices about four to seven yards wide. The first operation is called working in the "Whole Mine," and the second working in the "Broken."

Long Wall then may be defined as any system of working in which the seam is removed at one operation: Bord and Pillar as any system in which the seam is removed by two or more series of workings.

The Bord and Pillar and Long Wall systems of working are adapted to different circumstances, so that an exact comparison is impossible, though a general one may be made as follows :—

1. *Ventilation.*—In Long Wall the air enters by the main gate, and dividing into two splits, passes along the face, returning by roads on the extreme right and left. Nothing can be simpler than this arrangement; very little brattice is required, and the air, having the shortest possible distance to travel, acquires the least possible heat from the strata, a matter of great importance in deep mines, and also requires the least possible ventilating pressure (*i.e.*, less expenditure of money) to set it in motion.

In Bord and Pillar the air also enters by the central drift or Mother-gate bord, and divides into two splits; but, as the air has to be taken into each bord, it has a very much longer distance to travel, and a great deal of brattice is required.

On the other hand, should there be much gas, it can be isolated to the bord in which it is being given off in Bord and Pillar; whilst in Long Wall it will foul the whole face on the inbye side.

2. *Produce.*—In Long Wall all the seam may be extracted, and whilst the weight of the roof helps to break down the coal at the face, it does not rest upon it long enough to crush the coal. This, combined with the small amount (if any at all) of nicking and narrow work, tends to the production of the maximum of round coal.

In Bord and Pillar all the seam cannot be extracted, as some coal must always be left in stooks, and in addition, a portion of the pillars is often lost by falls of roof. In the whole workings, small is produced by nicking and narrow work and often in the broken by crush. The result being a smaller production both of unscreened and of round coal than in the Long Wall method of working.

3. *Cost.*—In Long Wall the cost of putting, supervision, and materials (*i.e.*, rails, sleepers, and brattice) will be less than in Bord and Pillar because the distance is shorter; and, as there is no yard work, and the weight of the roof helps to bring down the coal, the cost of hewing also will be less. On the other hand, shift and stone work will be very expensive; so much so, that where powder cannot be used, Long Wall is, in many cases, inadmissible.

A given length of face will stow more men in Long Wall than in Bord and Pillar.

4. *Surface Damage.*—When it is intended to work out the whole of the seam less damage is done by Long Wall than by Bord and Pillar, because the space formerly occupied by the seam is filled up by the stowage, and though this cannot be done so completely as to support the weight of the superincumbent strata without considerable compression of the stowage, yet the character of the support is the same over the whole area, and the surface is let down gradually and uniformly.

In Bord and Pillar the surface damage usually takes the form of irregular depressions dotted about here and there, putting a stop to all farm drainage.

5. *Accidents.*—Accidents from falls of stone are less likely to happen in Long Wall than in the broken workings of Bord and Pillar; and, as no coal is left below ground, under ground fires, from the spontaneous combustion of small coal crushed and ground together by falls of roof, are impossible. On the other hand, gas cannot be isolated to the place where it is being given off, as in Bord and Pillar. And in Long Wall, the men being closer together, should an explosion occur, more are likely to be killed.

Summary.—Long Wall is suitable for thin seams (less than four feet) or very thick (more than twelve feet) seams, lying at any angle; especially when they produce sufficient refuse for stowage and contain no gas and few troubles.

Bord and Pillar is suitable for seams of moderate thickness (from $3\frac{1}{2}$ to 8 feet) lying at low angles: especially if there be gas and troubles.

Stoping.

In the mining of metalliferous veins a Bord and Pillar system is adopted which is called *stoping*. It may be shortly described as follows:—

The vein having been cut up by means of levels and winzes into pillars, 25 to 50 yards in length by 15 to 30 yards in height, is worked by one or other of two methods, viz.:—

   1. By overhand stoping.
   2. By underhand stoping.

   1. *Overhand stoping.*—A jud is worked off (the full width of the vein and by 5—6 feet in height) starting from one of the lower corners of a pillar and carried right across the pillar horizontally from winze to winze. This jud, having gone 4 or 5 yards, is followed by another jud immediately above it; this is followed by a third, and so on. So that the portion of a pillar, still unworked, looks like a staircase, beneath which the miners stand; and the portion worked, which is stowed up with the deads, looks like a staircase upon which (or sometimes upon timbering) the miners stand. The useful mineral is separated from the deads and passed down from step to step, until it reaches the rolley-way below. Or else passages are left for it in the stowage down to the rolley-way level, with a sliding shutter in their lower ends by means of which the tubs are filled.

   2. *Underhand stoping.*—A jud is worked off, beginning at the upper corner of a pillar, and carried right across horizontally from winze to winze. When this jud has gone a few yards, a second is set away immediately below it, and so on, so that the unworked portion of the pillar is like a staircase, upon the steps of which the workmen stand. The useful mineral is separated from the deads and passes down the staircase from step to step, until it reaches the rolley-way level. The deads are stowed away on timber above the miners' heads.

   Comparison of the two methods.—The ore is broken down more cheaply by overhand than by underhand stoping; the leading of the useful mineral down the spouts is cheaper than passing it down the steps; and the stowage of the deads is more easily accomplished. The consumption of timber depends, perhaps, more upon the circumstances of the lode than upon the system adopted; but, as a rule, less will be required for overhand than for underhand stoping. The roof of unworked ore in overhand stoping, except when it is of a very friable nature, will be safer than the stowage roof in underhand stoping.

On the other hand, if the mineral be of a very valuable character portions of it may be lost in the stowage on its passage down the spouts or down the steps of stowage.

Shortly.—Overhand stoping is the most generally applicable. Underhand stoping—more costly from the expense of timbering, the greater difficulty of breaking the ore, of stowing the goaf, and leading away the useful mineral—is suited for those mines where the great value of the mineral makes the loss of a small quantity a matter of great importance, and for those where the ore is so friable as to make a dangerous roof for the working places. It is adopted in some German, a few English, and many South American mines.

Cost of Working.

This is very variable, depending upon the price of labour and the nature of the deposit.

In the case of coal, about one half of the labour cost is due to hewing, one-third to other underground labour, and the remaining one-sixth to surface labour. To this must be added materials, rents, rates, fuel, agency, depreciation, and interest on capital. In all, perhaps, about 5s. per ton on unscreened coal into waggons at the pit, in ordinary conditions of trade.

Collins in his book, referred to below, gives some costs of labour in metal mines.

The following books may be consulted:—"Metal Mining," Collins; "Metalliferous Minerals and Mining," Davies; "Ore Deposits," J. A. Phillips; Trans. N.E.I., vi. and vii.; and the books on Mining already mentioned.

WINDING.

Winding Engines.

The work to be done is not continuous for more than a few seconds, and is variable in amount, being greatest at the lift when the whole weight of the rope and the inertia of the mass set in motion have to be overcome. The weight of the rope is counterbalanced, as see below. The resistance due to the inertia of the load may be found by the following rule :—

  Let R = Resistance in lbs.
   W = Weight of load in lbs.
   V = Maximum velocity in feet per second
   g = Force of gravity = 32.
   T = Time in seconds taken to acquire the velocity V.

(111.) $R = \dfrac{WV}{gT}$ (See "Practical Mechanics," by Twisden, p. 231.)

The friction of the guides has to be overcome.

In designing a Winding engine to do any given work, consult the "Mechanical Engineering of Collieries," by Percy.

Counterbalances.—The common form in the north is the chain and staple. See the description of the Silksworth counterbalance. Trans. N.E.I., xxv. 201.

The Incline Counterbalance was adopted at Killingworth Colliery.

 F = Counterbalancing force in lbs. for a short distance on any portion of the incline.
 W = Weight of counterbalance in lbs.
 H = Height of the portion of the incline in feet.
 L = Length   .   do.   do.

(112.) $F = \dfrac{WH}{L} \therefore H = \dfrac{FL}{W}$

The Pendulum Counterbalance is used at Dudley Colliery.

F = Conterbalancing force in lbs. in any position of the pendulum.
W = Weight in lbs. of the counterbalance.
A = Angle the pendulum makes with the horizontal.

(113.) $F = \dfrac{W \cos. A}{\cos \dfrac{A}{2}}$

The Tail Rope Counterbalance as used at Garswood Colliery, near Wigan, consists of a rope of the same size as the winding rope. It passes round a pulley at the bottom of the shaft, and has one end fastened to the bottom of one cage, the other end to the bottom of the other cage.

The Koepe system does away with the winding drum altogether, and substitutes a sheave connected with the engine at bank. There is a return sheave at the bottom of the shaft. Two ropes are used, one connected with the tops of the cages and passing round the sheave at bank; the other connected with the bottoms of the cages, and passing round the sheave in the sump.

The Conical Drum, as used at Boldon—the full cage at the bottom of the shaft being attached to the small diameter of the cone, the empty cage at the top of the shaft being attached to the large diameter of the cone.

Automatic Variable Expansion is sometimes employed.

Drums.—Cylindrical, vertical (flat rope), or conical.

Pulley-Frames.

The principal strains are in two directions, viz., one vertical, due to the weight of the load; the other more or less horizontal, due to the pull of the engine. Timber, iron, and masonry will bear a crushing strain better than a tensile strain, or breaking across. In constructing pulley frames, therefore, the materials should be so placed as to be subjected to a crushing strain. This may be done in more than one way; but, in practice, it is found most convenient to employ two main struts, one vertical, parallel with the vertical portion of the rope; the other, more or less horizontal, parallel with the horizontal part of the rope.

In order to fix the size of long timber struts, see formula (41).

Pulleys.

The following rule is given for the diameter of round iron or steel rope pulleys, viz.:—

    Rope, 1 in. cir., requires pulley 10 ft. diameter.
      „ $1\frac{1}{4}$ in. „   „ $10\frac{1}{2}$ ft. „
      „ $1\frac{1}{2}$ in. „   „ 11 ft. „
      „ $1\frac{3}{4}$ in. „   „ $11\frac{1}{2}$ ft. „

and so on.

In order to save the ropes, the pulleys are sometimes set on springs, as at Cambois Colliery.

Let F = Force in lbs. applied at rim of pulley required to overcome friction of axle.
W = Weight upon pulley axle in lbs.
D = Dia. of pulley in inches.
d = Dia. of axle in inches.
m = Coef. of friction (say 0·07).

(114.) $$F = \frac{Wmd}{D}.$$

Ropes, Chains, Cages.

Ropes are either round, flat, or tapering, and are made of hemp, aloes, iron, or steel. For deep pits, round, steel ropes, are most in favour.

For the strength of pit ropes, see formulæ (1 to 21).

Chains should be annealed occasionally, otherwise they become brittle, and are likely to snap. The general rule in the north is to anneal cage chains once a month; annealing too often, decreases their tensile strength.

For strength of chains see formula (22).

Cages are made of iron or steel, have from one to four decks, and carry from one to eight tubs. An iron cage weighs about ⅔ of its load of full tubs; a steel cage about ⅓ of its load.

Sundries.

Detaching hooks.—See Trans. N.E.I., xxix., 201.

Safety cages are not much used in the north.

To find meetings, &c., with flat ropes.

Let n = Half the number of revolutions.
 d = Distance of meetings from bottom of pit in inches.
 r = Radius of drum at lift in inches + ½ t.
 t = Thickness of rope in inches.

(115.) $d = 3\cdot1416 n (2r + \overline{n-1} t)$.

Let D = Depth of pit in inches.
 n = Number of revolutions.
 r = Radius of drum at lift in inches + ½ t.
 t = Thickness of rope in inches.

$D = 3\cdot1416 n (2r + \overline{n-1} t.)$

(116.) $n = \sqrt{\left(\dfrac{r}{t} - \dfrac{1}{2}\right)^2 + \dfrac{D}{3\cdot1416 t}} - \left(\dfrac{r}{t} - \dfrac{1}{2}\right)$

(117.) $r = \dfrac{D - \{n(n-1)\, 3\cdot1416\, t\}}{2 \times 3\cdot1416 n}$.

The following books may be consulted:—André, Greenwell, and Percy, already mentioned; Trans., N.E.I., xxv.

DRAINING.

Pumps, &c.

The Lifting Pump has the engine situated at bank. Advantages and disadvantages of, viz.:—The engine is easily got at for repairs, engine cannot be drowned up, pumps can be carried down to a great depth, working parts even when drowned easily brought to bank for repairs. On the other hand: First cost very large, working cost large, take up much room in shaft.

The Forcing Pump may have its engine either at bank or in the mine. Advantages and disadvantages of, viz.:— First, when engine at bank: The engine is easily got at for repairs, engine cannot be drowned, pumps can be carried down to a great depth, the spears balance the column of water. On the other hand: First cost very large, working cost large, take up much room in shaft when more than one rising main required, and, if the working parts are drowned, they cannot be brought to bank for repairs. Second, when engine in pit: First cost is small, working cost small, take up little room in shaft. On the other hand: Danger of engine being drowned, engine not so easily got at for repairs, difficult to make joints in the rising main, and to make clacks to stand the pressure at a great depth.

The Syphon is used for bringing water over a ridge from a higher to a lower level. The short leg *must* not have a vertical length of more than 34 feet, *i.e.*, the ridge over which the water is to be lifted must not be more than 34 feet. (The greatest height of ridge in practice depends upon the special circumstances of each case.) The long leg *need* not have a vertical length of more than 34 feet. The effective pressure, expressed in feet of water column, is equal to the vertical length of the long leg in feet (not more than 34 feet), less the vertical length of the short leg in feet.

The Shaft.

1st. *The water tub* can be used where the quantity is small; but, in some cases, large quantities have been raised in this way.

2nd. *The Winding engine pump* is a pump attached to the winding engine, usually at night; but, sometimes, whilst coal is being drawn.

3rd. *The Lifting engine* is situated at bank, and may be either a Beam engine, or Rotative engine.

4th. *The Forcing engine* may be situated at bank and may be a Cornish engine, a Bull engine, a Rotative engine; or it may be situated in the mine, and be Direct-acting or Rotative.

The Workings.

1st. *Occurrence of feeders.*—The coal-field is basin-shaped, and formed of alternate layers of permeable and impermeable strata. Water, entering at the outcrop, will run through the first until intercepted by faults, fissures, pumping shafts, &c. In the North of England Coal-field shallow pits wet; deep pits dry.

When sinking, therefore, the feeders should be tubbed back in succession, as soon as they are encountered.

2nd. *Water levels and under level drifts.*—This is the best way of dealing with water, when possible.

3rd. *Raising water from the dip* by means of—The water-tub, the hand-pump, or horse-pump, when the units of work are not very great. The tail-rope pump, steam-engine, and spears, steam-engine and rope, compressed air, and electricity, when the units of work are large. Syphons, hydraulic engines and water-wheels, may be used in some special cases.

In dealing with constant feeders by men or horses: find the units of power by multiplying the lbs. of water per minute by the vertical distance in feet and add $50°/_0$ for friction. Then :—

1 man will be required per   900   units of power.
1 horse   „   6,000   „

The man (or horse) will not of course work continuously

for 24 hours per day; but if he works, say for 8 hours, *i.e.*, $\frac{1}{3}$ of a day, he will do $900 \times 3 = 2,700$ units of work per minute, which is equivalent to $\frac{2,700}{3} = 900$ units per minute of continuous work; and as the feeder is continuous, this is the most convenient way of making the calculation.

Sundries.

Acid water corrodes pumps, and its effects are enhanced by pressure. At Killingworth, the acid water was neutralised by a mixture of water and lime from magnesian limestone.

$$\text{Sulphuric Acid} + \text{Magnesian Limestone} = \text{Sulphate of Lime and Magnesia} + \text{Carbonic Acid} + \text{Water}.$$

(118.) $H_2SO_4 + (MgCa)CO_3 = (MgCa)SO_4 + CO_2 + H_2O$.

Red water pollutes streams. It may generally be cleared by heating. At Backworth, the red water was passed through the condenser.

$$\text{Soluble Hydrogen Carbonate of Iron} = \text{Insoluble Carbonate of Iron} + \text{Carbonic Acid} + \text{Water}.$$

(119.) $FeH_2(CO_3)_2 = FeCO_3 + CO_2 + H_2O$.

The Air Vessel is used to equalise the pressure, and so prevent shocks. As air, in contact with water, especially when under pressure, is absorbed by the water, it is necessary to supply the air vessel with fresh air. One of the most effective methods of doing this is to pump air in by means of an air-pump. See Trans. N.E.I., xxi.

Joints and Valves of special construction are required, to stand high pressures of water.

If D = Diameter of pump in inches.
G = Gallons per three-feet stroke.

(120.) $G = \frac{D^2}{10} \cdot (+2\%) \quad \therefore D = \sqrt{10G}$.

If D = Diameter of pump in inches.
L = Length of strokes in feet.
N = Number of strokes per minute.
G = Gallons delivered per minute.

(121.) $G = \cdot 034 LND^2 \therefore D = \sqrt{\dfrac{G}{\cdot 034 LN}}$

A pump delivers from 5 to 20% less than the actual amount due to diameter and length of stroke.

Davey gives the following rule for speed of his differential pump.

If L = Length of stroke of pump in feet.
S = Speed of pump piston, or ram, in feet per minute.

(122.) $S = \sqrt{L} \times 60.$

The speed of water in pipes should not exceed 200 to 250 feet per minute.

A convenient method of measuring feeders regularly is by means of a clay dam built across the drift. A thin iron plate is substituted for the top plank and in it a rectangular notch is cut through which the water flows.

If G = Gallons per minute.
d = Depth in inches of the sill of the notch below the surface of the water.
l = Length of notch in inches.

(123.) $G = 2 \cdot 67 ld \sqrt{d}.$

The depth d must be measured in still water, *i.e.*, two or three feet away from the notch.

If F = Depth of shaft in fathoms.
P = Pressure of water in lbs. per square inch.

(124.) $P = 2 \cdot 6 F.(-0 \cdot 8\%).$

If H = Head of water in feet required to overcome resistance.
G = Gallons per minute.
L = Length of pipe in yards.
D = Diameter of pipe in inches.

(125.) $H = \dfrac{G^2 L}{(3D)^5}$

For strength of pipes, see formula (23).
A pint of pure water weighs a pound and a quarter.
1 gallon = 277·25 cub. in. = 0·16 cub. ft. = 10 lbs.
400 gallons = 64 cub. ft., 1 cub. in. = 0·036 lbs., 1 cub. yard = 168·75 gallons.
1 cub. ft. = 6·25 gallons = 62·5 lbs. = 1,000 oz.
1 cub. fm. = 6 tons, 1 ton = 35·84 cub. ft.

The general cost of pumping from the mines of Northumberland and Durham, exclusive of interest and redemption of capital, is about one farthing per ton of water lifted 100 fathoms (Trans. N.E.I., xii., p. 181).

Boring against Old Workings.

Read the *Mines' Act General Rules, No.* 13.

The object of boring is to interpose a shell of solid coal between the exploring drifts and the old workings, which will act as a protection against water or gas.

Thickness of shell required depends upon the head of water in the old workings, the character of the seam, and the width of the drifts; but, from the want of experimental data upon the strength of coal, no very definite answer can be given to this question. In order to insure the thickness required a plan should be made, and the bore-holes set away accordingly.

Precautions to be adopted.—Bore-holes can be depended upon if made by a machine, for a distance not exceeding 30 feet. The Master Shifter should see that the holes are put up to their full distance each night, and the foreshift Deputy should measure them before permitting the hewers to begin work. The Overman and Back-Overman should see each day that the holes are being driven in accordance with the plan drawn up by the Certificated Manager. A mall and plugs should be kept ready in the face of the drift. Spare lamps should be placed a few yards outbye. When the head of water is very great, a tap, large enough to take the rods, may be wedged into the hole.

G = Gallons per hour.
L = Length of bore-hole in yards.
H = Head of water in feet.
D = Diameter of bore-hole in inches.

(126.) $G = \dfrac{\sqrt{(15D)^5 \, H}}{L}$

In calculating the time required to empty old workings (or any other reservoir) the fact that the value of H decreases must not be lost sight of. If the reservoir be parallel sided, assume H to remain unaltered and multiply the time thus obtained by 2. But if the reservoir be of irregular shape, divide it into thin horizontal laminæ and calculate the time required for the discharge of each separately.

Dams.

The object of a dam is to shut back water or gas.
Dams may be classed as—

1st. *Straight*, made of clay, brick, or wood.
2nd. *Wedge-shaped*, of wood.
3rd. *Cylindrical*, of brick or wood.
4th. *Spherical*, of brick or wood.

Straight brick dams are the commonest form in the north; but a spherical wooden dam is the strongest. Brick is liable to be cracked by a movement of the strata; and some Engineers advise that an india-rubber sheet should be placed upon it, next to the water.

For the strength of cylindrical and spherical dams, see formulæ (42) and (43).

Precautions to be adopted.—The spot chosen should be free from fissures, and not near any dislocation. The sides of the drift should be carefully hand-dressed. A pipe should be carried through the dam, near the top, to permit the air to escape.

Tubbing.

Tubbing may be segmental or cylindrical. For the fixing of tubbing, see either André or Greenwell.

For the strength of tubbing, see formulæ (42) to (44), and the following books may be consulted:—" Mine Engineering," Greenwell; " Practical Treatise on Coal Mining," André; Trans. N.E.I., xii., xv., xxi., xxiii., &c.

HAULING.

Resistances to be Overcome.

1st. *Friction* varies directly as the weight of the tub and the diameter of the axle; and inversely as the diameter of the wheel. It is but little affected by velocity (*i.e.*, the low velocity of mine haulage), but depends very much upon the state of the road (see Trans. N.E.I., vol. xxxii.). With ordinary wheels and axles it is about 50 lbs. per ton on a good macadamised road, 10 lbs. per ton on a railway, and 24 lbs. per ton on an underground rolleyway.

If F = Resistance in lbs. due to friction upon a level rolleyway, in fair condition.
W = Weight of tub in lbs.
D = Diameter of wheel in inches.
d =   ,,   of axle   ,,
m = Coefficient of friction = $0·0882 \dfrac{d}{D}$
    = $\left(\sin a - \dfrac{2L}{gT^2}\right) \cos a$ in (128).

(127.) F = mW.

In order to find by experiment the resistance of friction upon a level road, an incline must be chosen with as regular a gradient as possible, and the time a tub takes to descend under the influence of gravity accurately measured.

Then if L = Length of incline in feet.
H = Height of incline in feet.
T = Time of descent in seconds.
W = Weight of tub in lbs.
R = Friction in lbs.
a = angle of inclination of incline.
g = gravity, say 32.

(128.) $R = W\left(\sin a - \dfrac{2L}{gT^2}\right)\cos a.$

Roughly, the gradient being moderate.

(129.) $R = W\left(\dfrac{H}{L} - \dfrac{L}{16T^2}\right)$

2nd. *Inclination of road* will retard the set going up-hill and assist it going down.

If I = Resistance in lbs. due to inclination.
W = Weight of tub in lbs.
L = Length of incline in feet.
H = Height of incline in feet.

(130.) $I = \dfrac{WH}{L}$

It follows, then, that—neglecting the fact that the resistance due to friction is rather less on an incline than on a level—if

R = Resistance in lbs. due to friction and inclination.

(131.) $R = mW + \dfrac{WH}{L}$ (going up-hill).

(132.) $R = mW - \dfrac{WH}{L}$ (going down-hill).

In order to find the gradient rising inbye, at which the resistance of the full set coming out is equal to the resistance of the empty set going in,

Let W = Weight of full set in cwts.
w = ,, empty set ,,
m = Coefficient of friction.
G = Rate of gradient = $\dfrac{L}{H}$.

(133.) $G = \dfrac{W + w}{m(W - w)}$

3rd. *Curves.*—A greater resistance is met with in going round curves than on a straight road. The wheels, being fast upon their axles, must slide upon the rails, and the flanges of the wheels grind against the rails. Molesworth, page 160, gives the following formula:—

W = Weight of vehicle in cwts.
R = Resistance due to curve in cwts.
r = Mean radius of curve in feet.
D = Gauge of way in feet.
L = Wheel base in feet.
F = Coefficient of friction of wheels on rail = 0·1 to 0·27.

$$(134.) \quad R = \frac{WF(D+L)}{2r}$$

This formula (134) is intended for railway carriages, trucks, &c.; with the higher coefficient of friction (0·27), however, it appears to be applicable to mine tubs.

In order to resist the tendency to fly off the way at a curve the outer rail should be raised. This, however, is not necessary with main and tail rope haulage.

E = Elevation of outer rail in inches.
V = Velocity in feet per second.
D and r = as above.

$$(135.) \quad E = \frac{3DV^2}{8r}$$

4th. Influence of *fast* or *loose* wheels; *conical* or *flat* treads; *parallel* or *radial* axles.

Motors.

1st. *Men.*—Seven men are considered to be equal to one horse, and one man's wage in the North of England, including house and fire-coal, is about the same as the cost of a horse.

2nd. *Horses.*—The units of work that an average horse can do, during a ten hours' shift, depend upon the speed at which he is driven, and the state of the ventilation. He is most efficient at a low speed, two or three miles an hour, and in a well ventilated mine. In these circumstances he will do about 22,000 foot-lbs. per minute ($\frac{2}{3}$ of a mechanical horse, see table XIV., p. 38), which is equal to a tractive force of 125 lbs. exerted through a distance of 20 miles in 10 hours. In practice, the velocity will be greater and the distance less.

H

3rd. *Self-acting inclines.*—The weight of the full set has to overcome the friction of the full set, the empty set, the sheave or rope-roll, and rollers + the weight of the empty set and rope. This last (the weight of the rope) is a variable quantity, and is greatest at the start.

Let L = Length of incline in feet.
H = Height of incline in feet.
a = angle of incline.
F = Weight of full set in lbs.
E = Weight of empty set in lbs.
T = Time running in seconds.
g = gravity = 32.
R = Weight of rope in lbs.
S = Weight of rollers and sheave in lbs.
m = Coefficient of friction of tubs on level road.
m' = Coefficient of friction of rollers and sheave = about 0·03 on an average.
W = Weight in lbs. of the whole mass in motion.
P = Force in lbs. moving the sets.

$$P = F \sin a - \{ mF \cos a + mE \cos a + m'S + E \sin a + R \sin a \}$$

$$= F \sin a - \{ m \cos a (F + E) + m'S + \sin a (E + R) \}$$

But $\sin a = \dfrac{H}{L}$ and $\cos a$ practically = 1. Therefore—

(136.) $$P = \dfrac{FH}{L} - \left\{ m(F+E) + m'S + \dfrac{H}{L}(E+R) \right\}$$

(137.) $$T = \sqrt{\dfrac{2LW}{g\left\{\dfrac{FH}{L} - m(F+E) + m'S + \dfrac{H}{L}(E+R)\right\}}}$$

(138.) $$\dfrac{H}{L} = \dfrac{m(F+E) + m'S + \dfrac{W2L}{gT^2}}{F - (E+R)}.$$

Self-acting inclines are suitable for straight roads rising inbye one inch to the yard and upwards.

4th. *The main and tail-rope* system is suited for a narrow plane, having a regular gradient, and several branches.

5th. *Endless chain or rope* is suited for wide straight undulating planes, without any branches. The gradient, &c., may be calculated from formulæ (136) to (138).

6th. *The main rope* system is used on planes dipping inbye one inch to the yard and upwards. The weight of the empty set has to overcome the friction of the empty set, the rollers, sheave, and rope.

$$P = E \sin a - (m \cos a\, E + m'S + m'R),$$
and we get

$$(139.)\ P = \frac{EH}{L} - \left\{ mE + m'S + m'R \right\}$$

$$(140.)\ T = \sqrt{\frac{2LW}{g\left\{\frac{EH}{L} - (mE + m'S + m'R)\right\}}}$$

$$(141.)\ \frac{H}{L} = \frac{mE + m'S + m'R + \frac{W2L}{gT^2}}{E}.$$

7th. *Compressed air-engines,* stationary and locomotive. *Hydraulic engines* and *electric motors* are also used.

A tub will hold about 50 lbs. of unscreened coal per cubic foot. The exact quantity depends of course upon the specific gravity of the coal; but also upon the size of tub and the proportion of round coal, the larger the tub and the greater the percentage of round the greater the weight per cubic foot that it will hold.

The cost of haulage is very variable, depending upon the gradients, the length of road, and quantity led. Excluding interest on capital, the cost per ton of coal conveyed one mile is about :—

$\frac{3}{4}d.$, ordinary railroad with locomotives.

$6d.$, ordinary road with horses, excluding maintenance of road.

$\frac{1}{4}d.$ (?), canal with horses, excluding maintenance of canal. Canal with steam-tugs in some cases as low as $\frac{1}{100}d.$, excluding maintenance of canal.

1¼d., *level* railroad with horses.
2½d., underground *level* rolley-way with horses.
2d., ordinary underground rolley-way, rope or chain haulage.

If the quantity be small, the rolley-way be level or dipping slightly in favour of the load, and about half-a-mile or under in length, horses can compete favourably with mechanical haulage. But with large quantities, steep gradients, and long distances, mechanical haulage is cheapest.

The following books may be consulted:—Books by André, Callon, Greenwell, already mentioned. Trans. N.E.I., iii. and xvii. The use of steam for canal-boat propulsion, Manchester Assoc. of Eng., Jan. 1886; Proc. I.C.E. xxxi.

GENERAL PROPERTIES OF AIR AND GASES.

Air and gases may be defined as elastic fluids in contradistinction from liquids which are inelastic fluids.

The elasticity of the air is used to determine the ventilating pressure in a mine by means of the water-gauge.

Air and gases are ponderable, that is to say, they have weight; but the weight of a given volume depends upon its pressure and temperature.

Pressure :—The weight of a given volume of any gas varies as the pressure. In order to find the pressure of the air, we use the barometer.

The standard atmospheric pressure at 32 Fahr. and sea-level $= 29\cdot922$ in. mer. $= 14\cdot696$ lbs. per sq. in. $= 2,116$ lbs. per sq. ft. $= 26,213$ ft. of homogeneous air column $= 33\cdot9$ ft. of water column.

To reduce a barometer reading at any point above sea-level to the corresponding reading at sea-level, the following approximate rule is given by Mattieu Williams in "Science in Short Chapters":—

To the observed reading add $0\cdot1''$ for each :—
85 ft. up to 510 ft. that the point is above sea-level.
90 ft. from 510 to 1140 ft.
95 ft. from 1140 to 1900 ft.
100 ft. when above 1900 ft.

Thus $28''$ at a point 2000 ft. above sea-level $= 30\cdot2''$ at sea-level.

Correction for temperature :—Mercury expands about $0\cdot0001$ of its volume for each degree Fahr. To reduce, therefore, a reading at any temperature to the corresponding reading at the standard temperature of $32°$, subtract $\frac{1}{10,000}$ of the observed height for each degree above $32°$; or, if the temperature be below $32°$, add $\frac{1}{10,000}$ for each degree.

Depth of pits :

    If R = Reading of barometer at lower station.
        r =   ,,   at higher ,,
    T = Temperature Fahr. at lower station.
        t =   ,,   at higher ,,
    H = Difference of level in feet.

(142.) $H = 56{,}300\ (\text{Log. R} - \text{Log. r}) \left(1 + \dfrac{T+t}{900}\right).$

$\therefore \text{Log. R} = \dfrac{H}{56{,}300 \left(1 + \dfrac{T+t}{900}\right)} + \text{log. r}.$

More simply :

(143.) $H = 49{,}000 \left(\dfrac{R-r}{R+r}\right)\left(1 + \dfrac{T+t}{900}\right)$

$\therefore R = r \left\{ \dfrac{49{,}000\ (900 + T + t) + 900\ H}{49{,}000\ (900 + T + t) - 900\ H} \right\}$

Very roughly, the mercury rises 1 inch for each 150 fathoms of depth.

Temperature :—The weight of a given volume of any gas varies inversely as its absolute temperature. Absolute temperature = 459 + Fahr. temp.

To find the weight of a given volume of any gas at any known temperature and pressure, 459 cub. ft. of air at 0° Fahr. and bar. 1 in. weigh 1·3253 lbs. Therefore, if

    V = Volume of air in cub. ft.
    W = Weight in lbs.
    I = Barometer in ins.
    t = Temperature Fahr.

(144.) $W = \dfrac{1{\cdot}3253\ IV}{459 + t}.$

To find the weight of any other gas, multiply the weight of air by the specific gravity of the gas. See p. 104.

Gas in goaves.—It has been estimated that the air-space in a goaf is equal to about one-sixth of the volume of the coal extracted.

Absorption of gases by liquids and solids as of air by water in a pump.

Gases enclosed in the pores of coal must be distinguished from the gases that enter into the chemical composition of coal. Sundry analyses of these enclosed, or *occluded gases* as they are called, are given in the following table:—

TABLE XXIV.

GASES ENCLOSED IN THE PORES OF COAL AND EVOLVED IN VACUO AT 212° FAHR.

Name of Colliery.	Quality.	CO_2.	O.	CH_4.	N.	Quantity CC per 100 Grams.	Cubic Feet per Ton.
Navigation	Steam	13·21	0·49	81·64	4·66	250	90
Dunraven	do.	5·46	0·44	84·22	9·88	218	78
Cyfarthfa	do.	18·90	1·02	67·47	12·61	147	52
Bute	do.	9·25	0·34	86·92	3·49	375	135
Bonville's Court	Anthracite	2·62	...	93·13	4·25	555	199
Watney's	do.	14·72	...	84·18	1·10	600	216
Plymouth Iron Works	Bituminous.	36·42	0·80	...	62·78	55·9	20
Cwm Clydach	do.	5·44	1·05	63·76	29·75	55·1	19·8
Bettwys	do.	22·16	6·09	2·68	69·07	24·0	8·6

(Thomas.)

Experiments of Mr. Lindsay Wood on the pressure of gases enclosed in coal. (See Trans. N.E.I., xxx.) The greatest pressure obtained was at Boldon, 461 lbs. per square inch.

TABLE XXV.

Transpiration of Gases.

That is to say, the passage of gases through minute tubes, such as the pores of coal.

Name of Gas.	Times for Transpiration of equal Volumes.	Velocities of Transpiration.
Oxygen (O)	1·000	1·000
Air	0·9030	1·1074
Nitrogen (N)	0·8768	1·141
Carbonic Oxide (CO)	0·8737	1·145
Carbonic Acid (CO_2)	0·7300	1·370
Marsh Gas (CH_4)	0·5510	1·815
Ethylene (C_2H_4)	0·5051	1·980
Hydrogen (H)	0·4370	2·288

(Graham.)

Practical bearing.—Gases flow from green coal into workings. Blowers. Gases assist hewer by breaking down coal.

The Diffusion of Gases.

When two gaseous bodies are mixed together they gradually diffuse themselves through each other; so that, after sufficient time has elapsed for the purpose, whatever may have been their relative densities, they are found intimately blended; the heavier gas does not fall to the bottom, nor does the lighter one rise to the top.

TABLE XXVI.

RELATIVE VELOCITY OF DIFFUSION.

—	Spg.	$\sqrt{Spg.}$	$\sqrt{\dfrac{1}{Spg.}}$	Velocity of Diffusion, Air being taken as unity.
Air	1·000	1·000	1·000	1·000
Hydrogen (H)	0·06926	0·2632	3·7794	3·83
Marsh Gas (CH_4)	0·559	0·7476	1·3375	1·344
Steam (H_2O)	0·6235	0·7896	1·2664	...
Carbonic Oxide (CO)	0·9678	0·9837	1·0165	1·0149
Nitrogen (N)	0·9713	0·9856	1·0147	1·0143
Ethylene (C_2H_4)	0·978	0·9889	1·0112	1·0191
Oxygen (O)	1·1056	1·0515	0·9510	0·9487
Sulphuretted Hydrogen (H_2S)	1·1912	1·0914	0·9162	0·95
Carbonic Acid (CO_2)	1·529	1·2365	0·8087	0·812

(Graham.)

The above table shows that fire-damp mixes with air more readily than stythe does; and fire-damp, therefore, is more easily cleared away by the ventilating current than stythe is.

TABLE XXVII.

FIRE-DAMP ANALYSES.

Name of Colliery.	CH_4.	N.	O.	CO_2.	H.	
Wallsend, from pipe on surface............	92·8	6·9	0·0	0·3	0·0	100·0
Jarrow, Bensham Seam	83·1	14·2	0·6	2·1	0·0	100·0
Hebburn, Do.	86·0	12·3	0·0	1·7	0·0	100·0
Jarrow, Low Main Seam	79·7	14·3	3·0	2·0	0·3	99·3
Jarrow, 5/4 Seam.........	93·4	4·9	0·0	1·7	0·0	100·0
Oakwellgate, Do.	98·2	1·3	0·0	0·5	0·0	100·0
Hebburn, Coal 24 ft. below Bensham ...	92·7	6·4	0·0	0·9	0·0	100·0

(De La Beche and Lyon Playfair.)

Authorities :—" Coal, Mine Gases, and Ventilation," by Thomas. Trans. N.E.I., xxx. Ganot's " Elementary Physics." " Practical Treatise on Gases met with in Coal Mines," by Atkinson. " Practical Treatise on Heat," by Box.

CHEMISTRY.

Compounds and Elements.

Substances may be divided into three classes.

(1.) Chemical compounds—those substances which can be split up by chemical processes into two or more different materials.

(2.) Chemical elements or simple substances—those which have hitherto resisted all attempts to split them up into two or more different materials. There are at present about 63 of these bodies.

(3.) Mechanical mixtures—substances formed from a mixture of the above.

Atoms.

The atomic *theory* has been adopted to explain the *fact*, that in chemical combinations elements unite in fixed proportions.

An atom is the smallest particle of an element that can *enter into chemical combination* with other elements.

Atoms are incapable of being divided.

The atoms of the same substance are similar to one another and equal in weight.

The atoms of different substances differ in weight.

The weight of the atom of hydrogen being taken as the unit; the atom of oxygen weighs 16, the atom of nitrogen 14, and so on.

Chemical Symbols.

The atoms of the elements are represented by symbols; the first letter of the name being generally taken to express the atom.

Thus, the atom of Oxygen is denoted by O.
   „   Nitrogen   „   N.
   „   Hydrogen   „   H, etc.

These symbols represent definite weights of the respective elements. H represents the unit of atomic weight, *i.e.*, the weight of the hydrogen atom, whatever that may be.

O represents a weight of Oxygen = 16 Hydrogen atoms.
N   „   Nitrogen = 14   „
C   „   Carbon   = 12   „

The symbols and atomic weights of the elements we are interested in are given in the following table :—

TABLE XXVIII.

SYMBOLS AND ATOMIC WEIGHTS.

Name of Element.	Symbol.	Atomic Weight.
Oxygen	O	16
Hydrogen	H	1
Nitrogen	N	14
Carbon	C	12
Sulphur	S	32
Phosphorus	P	31
Chlorine	Cl	35·5
Potassium	K	39
Sodium	Na	23
Calcium	Ca	40
Manganese	Mn	55
Magnesium	Mg	24
Iron	Fe	56
Zinc	Zn	65

Molecules and Formulæ.

The group of atoms forming the smallest particle of a compound which can *exist in a free state*, is called its molecule; and the molecule of a compound is expressed by putting together the symbols of the atoms which compose it. This group of symbols is called a formula.

Thus the molecule of water contains one atom of oxygen, and two atoms of hydrogen, and may, therefore, be expressed by the formula HHO. When, however, several similar atoms are present, the symbol is only written once, and a small number is put on the *right* of it, and *a little below*, to show how many atoms are present.

Thus the formula for the molecule of water is H_2O.

When more than one molecule has to be represented a number is placed on *the left* and *level*.

Thus four molecules of water are represented by $4H_2O$.

The molecule of many of the elements consists of two atoms.

Chemical Equations.

Chemical changes are represented by equations.

Thus, $Zn + H_2SO_4 = H_2 + ZnSO_4$ signifies that 65 parts by weight of zinc reacting on 98 parts by weight of sulphuric acid, form 2 parts by weight of hydrogen and 161 parts by weight of zinc sulphate.

Equal volumes of all gases contain, under the same conditions, the same number of molecules; equations, therefore, representing changes in which gases take part, may be read off at once in volumes.

If the volume occupied by one atom of Hydrogen be taken as unity, one molecule of each of the gases will occupy two such volumes. Thus:

$$CH_4 + 2O_2 = CO_2 + 2H_2O$$

may be read :—

Two volumes of marsh gas and four volumes of oxygen, form two volumes of carbonic acid gas and four volumes of vapour of water.

The following books may be consulted : " Inorganic Chemistry," W. A. Miller ; " Exercises in Practical Chemistry," Harcourt and Madan ; " The New Chemistry." International Science Series.

THE GASES.

Oxygen.

Symbol, O ; atomic weight, 16.

1,000 cubic feet at 32° Fahr. and bar. 30 in. weigh 89·342 lbs.

Oxygen forms by weight $\frac{8}{9}$ of water, $\frac{1}{4}$ of the atmosphere, and $\frac{1}{2}$ of the solid crust of the earth. It was discovered by Priestley in 1774 ; and has neither colour, taste nor smell. Oxygen is occasionally found amongst the occluded gases; but principally occurs in mines as a constituent of air. It is essential to life ; but, undiluted, it is not fit to be breathed for more than a short time. It supports combustion, and substances which burn in air burn fiercely in oxygen.

It may be prepared from a mixture of potassium chlorate four parts and manganese dioxide one part, mixed together and heated. The whole of the oxygen contained in the potassium chlorate is given off, and a compound of potassium and chlorine remains.

(145.) $$\text{Potassium Chlorate} = \text{Potassium Chloride} + \text{Oxygen.}$$
$$2KClO_3 = 2KCl + 3O_2.$$

The manganese dioxide is unaltered ; in fact, the oxygen could be obtained from potassium chlorate alone; but it is found in practice that the presence of manganese dioxide materially assists the operation.

Carbonic Oxide.

Formula, CO ; molecular weight, 28.

1,000 cubic feet at 32° Fahr. and bar. 30 in. weigh 78·305 lbs.

This gas is the result of imperfect combustion. When a body containing carbon is burnt in air, each atom of carbon will combine with two atoms of oxygen to form carbonic acid gas; but, if there is not sufficient air to provide two atoms of oxygen for each atom of carbon, that is to say, if the combustion of the carbon is incomplete, carbonic oxide is formed. It has been detected in rare cases amongst the occluded gases; and is also produced by the combustion of coke, charcoal, and gunpowder; and must, in many cases, be one of the constituents of after-damp.

It has neither colour, taste, nor smell, but is exceedingly poisonous; $\frac{1}{2}$ per cent. in the air, if breathed for long, producing fatal results. It does not support combustion, but itself burns with a blue flame, forming CO_2.

It may be prepared from hydrogen oxalate, treated with hydrogen sulphate. Carbonic oxide and carbonic acid are driven off, the latter of which is removed by passing the mixture through a solution of potassium hydrate; but, as this gas is very poisonous, it is best not meddled with by unskilled persons.

$$\text{Hydrogen Oxalate} + \text{Hydrogen Sulphate} = \text{Carbonic Oxide} + \text{Carbonic Acid} + \text{Water} + \text{Hydrogen Sulphate}.$$

(146.) $H_2C_2O_4 + H_2SO_4 = CO + CO_2 + H_2O + H_2SO_4$

Hydrogen.

Symbol, H; atomic weight, 1.

1,000 cubic feet at 32° Fahr. and bar. 30 in. weigh 5·5832 lbs.

Hydrogen has neither colour, taste, nor smell. It is very inflammable, burning with an almost colourless flame. If breathed in its undiluted state, it quickly causes a very disagreeable sensation; but this is due to the exclusion of oxygen from the lungs, and not to the properties of hydrogen, which is not poisonous, and may be breathed when diluted with ten times its volume of air, for a considerable time, without experiencing any ill effect.

The experiments of Meyer and Thomas show that, in an explosion of marsh gas and air, the whole of the marsh gas is broken up; and, if there be too little air to form carbonic acid gas and water, carbonic oxide and hydrogen are formed.

It may be prepared by treating zinc with hydrogen sulphate. The hydrogen is driven off, and zinc sulphate is left behind.

$$\text{Zinc} + \text{Hydrogen Sulphate} = \text{Hydrogen} + \text{Zinc Sulphate.}$$

(147.) $\quad \text{Zn} + \text{H}_2\text{SO}_4 = \text{H}_2 + \text{ZnSO}_4.$

Combined with carbon in the proportion of 4 parts by weight of hydrogen to 12 of carbon, it forms marsh gas, the principal constituent of fire-damp.

Hydrogen Sulphide.

Formula, H_2S; molecular weight, 34.

1,000 cubic feet at 32° Fahr. and bar. 30 in. weigh 94·92 lbs.

Hydrogen sulphide, or sulphuretted hydrogen as it is more generally called, is a colourless gas, but has a strong smell not unlike that of rotten eggs. It is "generated in small quantity in coal mines, more especially in old-worked portions, which are partly filled with water. By the action of oxygen dissolved in water, sulphates are formed; props in undergoing decomposition in water break up the sulphate of lime and assimilate its oxygen, the sulphur seizing probably the hydrogen of the wood to form hydrogen sulphide." (Thomas's "Coal, Mine Gases, and Ventilation," p. 204.)

It does not support combustion, but is itself inflammable, forming water and sulphurous anhydride ($H_2O + SO_2$). Breathed in an undiluted state, it is fatal to life; and, when diluted with ten times its volume of air, it produces sickness, giddiness, weakness, and loss of sensation.

This gas may be prepared from proto-sulphide of iron treated with dilute hydrogen chloride. Hydrogen sulphide will be given off, and iron chloride formed.

$$\text{Iron Proto Sulphide} + \text{Hydrogen Chloride} = \text{Iron Chloride} + \text{Hydrogen Sulphide.}$$

(148.) $\quad \text{FeS} + 2\text{HCl} = \text{FeCl}_2 + \text{H}_2\text{S}.$

Nitrogen.

Symbol, N; atomic weight, 14.

1,000 cubic feet at 32° Fahr. and bar. 30 in. weigh 78·175 lbs.

Nitrogen has neither colour, taste, nor smell, and is incapable of supporting combustion or animal life, but is not poisonous, causing death when breathed only by excluding oxygen from the lungs. It is found in large quantities amongst the gases occluded in some coals; but occurs principally in mines as a constituent of air.

Mixed with oxygen, it forms air, and the readiest way of obtaining it for experiment is to withdraw the oxygen by the action of some substance which has an affinity for oxygen and not for nitrogen. Phosphorus is convenient for this purpose, since it readily combines with oxygen; and the compound formed, phosphorus pentoxide (P_2O_5), is soluble in water, and is, therefore, quickly absorbed when the experiment is made over the pneumatic trough, leaving the nitrogen nearly pure.

Carbonic Acid Gas.

Formula, CO_2; molecular weight, 44.

1,000 cubic feet at 32° Fahr. and bar. 30 in. weigh 128·45 lbs.

Carbonic acid gas has neither colour nor smell, but an acid taste. It is found in large quantities amongst the gases occluded in some coals; but is also produced in mines by the respiration of men and animals, by the burning of candles and lamps, and by the oxidation of the coal and other substances.

It extinguishes lights, and is fatal to animal life.

It may be prepared by the decomposition of marble by hydrogen chloride. Carbonic acid gas is given off, calcium chloride and water are formed in the vessel.

$$\text{Marble} + \text{Hydrogen Chloride} = \text{Carbonic Acid Gas} + \text{Calcium Chloride} + \text{Water.}$$
$$(149.)\ CaCO_3 + 2HCl = CO_2 + CaCl_2 + H_2O$$

Fire-Damp.

Marsh gas. Formula, CH_4; molecular weight, 16.

1,000 cubic feet at 32° Fahr. and bar. 30 in. weigh 45·22 lbs.

Fire-damp is a mixture of several gases, its principal constituent being marsh gas, CH_4; but its composition varies at different collieries, as see p. 105; and, in addition to these gases, coal-dust is often present. It is only found in mines as an occluded gas.

Marsh gas may be prepared by heating a mixture of sodic

acetate and sodic hydrate in an iron tube. Marsh gas is driven off and sodic carbonate is formed in the tube.

$$\text{Sodic Acetate} + \text{Sodic Hydrate} = \text{Sodic Carbonate} + \text{Marsh Gas.}$$

(150.) $NaC_2H_3O_2 + NaHO = Na_2CO_3 + CH_4.$

For making experiments, ordinary coal gas may be used, though it differs in composition from average specimens of fire-damp.

TABLE XXIX.

COMPOSITION OF FIRE-DAMP AND COAL GAS.

	$CH_4.$	H.	CO.	$CO_2.$	$C_2H_4.$	N.
Fire-damp...	94·0	0	0	1·0	0	5·0
Coal Gas ...	42·0	42·0	4·5	0	9·0	2·5

The exact effects of fire-damp upon combustion and animal life depend upon its composition, temperature, and density; but, speaking generally, at ordinary temperatures and pressures, when mixed with 3·5 times its volume of air, it does not explode, but burns quietly; with 5·5 volumes of air, it explodes slightly; and with about $9\frac{1}{2}$ volumes of air the explosion is the greatest. With 13 volumes of air, it explodes feebly; with 30 volumes of air, it will show plainly on the lamp; with 50 volumes of air, it can just be detected on the lamp by a skilful observer. If breathed in an undiluted state, it would soon cause death; but, mixed with twice its own volume of air, it may be breathed for some time without ill effects.

An explosion of marsh gas and air by volume.

(1) Suppose we have of CH_4 2 volumes.
(2) C requires $O_2 = 2$ volumes of oxygen forming 2 volumes of CO_2.
H_4 also requires $O_2 = 2$ volumes of oxygen forming 4 volumes of H_2O.

(3) $CH_4 \therefore$ requires 4 volumes of oxygen. But 1 volume of air contains ·21 volumes of oxygen.
∴ 19 volumes of air will be required for 4 volumes of oxygen.
∴ 19 volumes of air are required for 2 volumes of CH_4.
∴ $9\frac{1}{2}$ volumes of air are required for 1 volume of CH_4.
And the composition of the after-damp is 1 volume CO_2 + 2 volumes H_2O (steam) + $7\frac{1}{2}$ volumes N.

In practice, the exact composition of the after-damp will depend upon the composition of the explosive mixture. In every case the whole of the marsh gas will be broken up and—if there be insufficient oxygen to consume all the marsh gas and coal dust, if the last be present—some carbonic oxide and hydrogen will be formed. In all probability, carbonic oxide is formed in the majority of explosions in mines. (Thomas's " Coal, Mine Gases, and Ventilation," p. 323.)

The force developed by an explosion of marsh gas and air depends upon a multitude of circumstances, many of which cannot be determined in the case of an explosion in a mine. But in the case of a mixture of marsh gas and air in the most explosive proportions, and enclosed in a strong vessel, we can calculate the force developed, as follows:—

1 lb. of CH_4 burning to CO_2 and H_2O yields about 23,550 units of heat. (See p. 34.)
Let the initial temperature be 62° Fahr. = 521° absolute.
 1 lb. CH_4 = 12 oz. C + 4 oz. H.
 12 oz. C + 32 oz. O = 2·75 lbs. CO_2.
 4 oz. H + 32 oz. O = 2·25 lbs. H_2O.
 And 64 oz. O are contained in about 17 lbs. air.

We have then, taking specific heat at constant volume. (See p. 33.)

CO_2	2·75 × ·1711 = ·470	units of heat to raise	2·75 lbs. CO_2 one deg.
H_2O	2·25 × ·364 = ·819	do.	2·25 lbs. H_2O one deg.
N	13·00 × ·1727 = 2·245	do.	13·00 lbs. N one deg.
	18·00 3·534		18·00

Then the degrees the mixture will be raised are $\frac{23550}{3\cdot534}$ = 6663° Fahr. and the volume it will seek to attain = $\frac{521+6663}{521}$ = 13·8, *i.e.*, the steady pressure due to the explosion = 13·8 atmospheres; but to this must be added a considerably increased force due to shock, the amount of which cannot be calculated. When it is remembered that 13·8 atmospheres are equal to 30,000 lbs. per sq. ft., whereas the force of a hurricane moving at the rate of 100·miles an hour is only 50 lbs. per sq. ft., some idea of the terrific force of an explosion may be realised.

At the Haswell Explosion, in 1844, Faraday and Lyell drew attention to the part that coal dust might play in an explosion; but, though the question has cropped up from time to time since, it is only recently that the matter has been thoroughly investigated. Mr. Galloway has, within the last few years, proved that an explosion may be caused with coal dust and air, without the presence of any gas. (See Trans. Royal Soc. 1876—1884, and "Nature," 6th Nov., 1884.) And it is now admitted that many of the most violent explosions of recent years were due in part, if not entirely, to coal dust. (See "Explosions in Coal Mines," by Atkinson.)

The detection of fire-damp is usually effected with the Davy lamp; but, since it has been shown by Sir Frederick Abel that, if coal dust be present, 1·5 of gas in the air will render the mixture explosive, a more delicate test is required. The following detectors are described in the Trans. N.E.I., viz.:—Ansell's, vol. xv.; Steavenson's, vol. xxvi.; Liveing's vol. xxvii.; Forbe's, vol. xxix.; and Maurice's, vol. xxxvi. Chatellier has introduced a lamp with screen and two shields for the same purpose.

Methods of dealing with Fire-damp.

Removal by :—Firing, now no longer practised. Drainage of goaves by pipes to the upcast (see Faraday and Lyell's report on the Haswell Explosion), or by bore-holes to the surface; both impracticable. Drainage of whole coal by bore-holes and gas drifts in a higher seam might answer in certain cases.

Dilution with air. (See General Rules 1.)

Absence of heat :—Heat is required for light and for shot-firing. The steel mill of Spedding about 1740. Reflected light, and fish-skins have been tried.

The safety lamp due to Clanny, 1811 ; Davy and Stephenson, 1815. The safety of the lamp depends upon the fact that metal gauze permits air and light to pass but not flame. The conducting power of the gauze is impaired by overheating, broken wires, dirt, or exposure to a current of gas.

The maximum of safety, combined with the maximum of light, is obtained from gauze with wires $\frac{1}{60}$ to $\frac{1}{40}$ in. dia., and spaced with 28 apertures to the linear inch.

The best-known lamps are the Davy, Geordie, Clanny, Mueseler, Tin-can, and Marsaut lamps.

The Accidents in Mines Commission (1886) report most favourably of the following :—Gray's, Marsaut's, the Bonneted Mueseler, and Evan Thomas's modification of the bonneted Clanny.

Swan's portable electric miners' lamp gives a good and perfectly safe light ; but it is too costly for practical use, and will not indicate the presence of carbonic acid gas.*

In the Author's opinion, too much attention is paid to securing lamps from the action of violent currents to which they are very unlikely ever to be exposed ; whilst their illuminating power, a daily necessity, is too much neglected. Inventors appear to forget the importance of a good light all round, including the roof. About $42°/_o$ of the fatal accidents in mines are due to falls of roof and sides, as against $24°/_o$ due to explosions, and a very small proportion of these last have been traced to the safety-lamp.

Shot-firing has already been dealt with under the head of explosives.

Ambulance Classes.

The St. John's Ambulance Association have established classes at many colliery villages, with excellent results. The

---

* It is described, together with the fire-damp detector attached, in Trans. N.E.I. xxxv.

following are short and clear instructions for the recovery of persons suffocated in mines.

Asphyxia.

Miners are exposed to asphyxia when the circulation of the air is not sufficiently active, when the mine exhales a quantity of deleterious gas, when they imprudently penetrate into ancient and abandoned workings, and when there is an explosion.

The symptoms of asphyxia are sudden cessation of the respiration, of the pulsations of the heart, and of the action of the senses; the countenance is swollen, and marked with reddish spots, the eyes are protruded, the features are distorted, and the face is often livid, &c.

The best and first remedy to employ, and in which the greatest confidence ought to be placed, is the renewal of the air necessary for respiration. In succession:—

1. Promptly withdraw the asphyxiated person from the deleterious place, and expose him to pure air.

2. Loosen the clothes round the neck and chest; and dash cold water in the face and on the chest.

3. Attempts should be made to irritate the pituitary membrane with the feathered end of a quill, which should be gently moved in the nostrils of the insensible person, or to stimulate it, with a bottle of volatile alkali placed under the nose.

4. Keep up the warmth of the body, and apply mustard plasters over the heart and round the ankles.

5. If these means fail to produce respiration Dr. Sylvester's method of producing artificial respiration should be tried, as follows:—

Place the patient on the back on a flat surface inclined a little upwards from the feet; raise and support the head and shoulders on a small firm cushion or folded article of dress placed under the shoulder-blades. Draw forward the patient's tongue and keep it projecting beyond the lips; an elastic band over the tongue and under the chin will answer this purpose, or a piece of string or tape may be tied round them, or by raising the lower jaw the teeth may be made to retain the tongue in that position. Remove all tight clothing from about the neck and chest, especially the braces. Then standing at the patient's head, grasp the arms just above the elbows, and draw the arms gently and steadily upwards above the head, and keep them stretched upwards for two seconds (by this means air is drawn into the lungs). Then turn down the patient's arms and press them gently and firmly for two seconds against the sides of the chest (by this means air is pressed out of the lungs). Repeat those measures alternately, deliberately, and perseveringly about fifteen times in a minute, until a spontaneous effort to respire is perceived; immediately upon which cease to imitate the movements of breathing, and proceed to induce circulation and warmth.

6. To promote warmth and circulation rub the limbs upwards with firm grasping pressure and energy using handkerchiefs, flannels, &c. Apply hot flannels, bottles of hot water, heated bricks, &c., to the pit of the stomach, the arm pits, between the thighs, and to the soles of the feet.

7. On the restoration of life a teaspoonful of warm water should be given; and then, if the power of swallowing has returned, small quantities of wine, warm brandy and water, or coffee should be administered.

8. These remedies should be promptly applied, and, as death does not certainly appear for a long time, they ought only to be discontinued when it is clearly confirmed. Absence of the pulsation of the heart is not a sure sign of death, neither is the want of respiration.

Illuminating Gas.

Illuminating gas is not found in mines; but, as it is an important product of the distillation of coal and is largely used about collieries, it will be as well to say a word about it here.

Its specific gravity depends upon the proportion of the heavier hydrocarbons present; in other words, specific gravity is a test of illuminating power.

12 candles gas is about 0·405 specific gravity.
14   „   „   0·430   „   „
16   „   „   0·455   „   „
18   „   „   0·482   „   „
20   „   „   0·508   „   „
22   „   „   0·537   „   „

Its composition is variable, that of an average specimen is given on p. 113.

The volume produced from a ton of coal depends upon the composition of the coal, as see Table XXX.

TABLE XXX.
AVERAGE PRODUCE OF A TON OF COAL.

	Newcastle Coal.	Wigan Coal.	Wigan Cannel.
Gas, cubic feet	9,500	9,980	10,900
Candle power	13	11·4	21·25
Coke, lbs.	1,540	1,517	1,436
Tar, gallons	9	11	17
Ammoniacal liquor, gallons	10	20	18

("Gas Manager's Pocket Book.")

Air.

1,000 cubic feet of air at temp. 32° Fahr. and 30 in. bar. weigh 80·9 lbs.

Air is a mixture of oxygen and nitrogen, with traces of some other gases, and vapour of water. Its composition varies a little in different places.

100 cubic feet of pure dry air contain :—
- Oxygen . . . . . 20·99 cub. ft.
- Nitrogen . . . . . 78·98   „
- Carbonic Acid Gas . . 0·03   „

　　　　　　　　　　100·00 cub. ft.

(Angus Smith.)

100 cubic feet of ordinary air contain :—
  Oxygen . . . . 20·61 cub. ft.
  Nitrogen . . . . . 77·95 ,,
  Carbonic Acid . . . 0·04 ,,
  Vapour of Water . . . 1·40 ,,

  100·00 cub. ft.

  Ammonia . . . . ⎱
  Nitric Acid . . . . ⎬ Traces.
  Carburetted Hydrogen . ⎰
  Sulphuretted Hydrogen . ⎱ Traces in large towns.
  Sulphur Dioxide . . . ⎰

Besides these gases, minute particles of solid matter are always present.

Sundry Analyses by Angus Smith :—

LONDON.
  Most Oxygen, Belsize Park . 21·010 per cent.
  Least Oxygen, Lambeth . . 20·795 ,,
  Badly Ventilated Law Court . 20·49 ,,
  Least CO_2 (open spaces) . . 0·0334 ,,
  Most CO_2 (gallery of theatre) . 0·32 ,,

METROPOLITAN RAILWAY.
  Oxygen... 20·7...Carbonic Acid . 0·1452 ,,

MINES.

	Average of 339 Mines.	Worst of 339 Mines.	Average of 8 Coal Mines.	Worst of 8 Coal Mines.
O ...	20·26	18·27	20·74	20·44
CO_2 ...	0·786	2·73	0·24	0·42

Analyses of air in the returns of 9 Mines in Saxony, by Dr. Winkler :—

  N.   75·6174 to 78·565 % by vol.
  O.   17·751 to 19·689   ,,
  CO_2. 0·1168 to 2·716   ,,
  H_2O. 2·5254 to 4·1904 ,,
  CH_4. 0·01754 to 0·25576 ,,

(N.E.I., xxxii., abs. p. 25.)

Purification by:—Diffusion, the wind, rain, plants, and animals.

Vitiation by:—Withdrawal of oxygen, as in the breathing of men and horses; the combustion of lamps, candles, and gunpowder; the conversion of iron pyrites into sulphate of iron, $2FeS_2 + 4O_2 = 2FeSO_4 + S_2$; the oxidation of small coal; &c., &c. And Vitiation by the introduction of foreign substances, such as the occluded gases, the products of combustion and respiration, vapour of water, and coal dust.

Angus Smith considers that two miners, using $\frac{1}{2}$ lb. candle and 12 oz. of powder, will produce $25\frac{1}{2}$ cub. ft. of CO_2 in 8 hours. Thomas, quoting Boussingault, says that a horse will produce 155 cub. ft. of CO_2 in 24 hours. But the quantity, both in men and horses, is very variable, and, as certain organic exhalations are also given off in breathing which, though they may not be capable of detection by chemical analysis, are more deleterious than CO_2, no estimate of the volume of air required can be formed from an attempt to calculate the CO_2 produced.

The quantity of air required depends upon the conditions of each mine. In the North of England, the volume seems to vary from 100 to 500 cubic feet per min. per person employed, and from 30 to 160 cubic feet per min. per ton of coal worked per day. The velocity in the workings should be about 4 feet per second.

Authorities:—" Air and Rain," Angus Smith; The Chesterfield and Derbyshire Inst. Trans., x.; "Coal and its Uses," Green, Miall, etc.; Papers by Davy, communicated to the Royal Soc. (See Royal Soc. Trans., 1816.) The Report of the Select Com., 1835; The Journal of the British Soc. of Mining Students, vi.; J. J. Atkinson on "Ventilation;" also, Papers by the same author in the Transactions of the North of England Institute; "Ventilation of Coal Mines," Fairley; "Coal, Mine Gases, and Ventilation," Thomas; J. Wales' Papers, Trans. N.E.I., vi. and vii.; "Hist. of Coal Mining," Galloway; Proceedings Royal Soc., xxiv. to xxxvii.; Report of the Accidents in Mines Commission, 1886; "Explosions in Coal Mines," Atkinson; "Ambulance Lectures," Weatherly; and "Shepherd's First Aid to the Injured," Bruce.

VENTILATION.

A WIND, either upon the surface or in the mine, results from a difference of pressure; the air passing from the place where the pressure is high to the place where the pressure is low.

If H = Height in feet of a column of air of the density of the flowing air that will, by its weight, produce the difference of pressure.

V = Velocity of the wind in feet per second.

(151.) $H = \dfrac{V^2}{64} \therefore V = 8\sqrt{H}.$

This formula applies to all fluids, H being taken in feet of the fluid in question.

In practice the actual velocity of an air current is much retarded by friction; and it is, therefore, necessary to study the laws that govern the resistance that air meets with in mines.

The Three Laws of Friction.*

(1) The pressure required to overcome the friction of the air increases and decreases in exactly the same proportion that the area or extent of the rubbing surface, exposed to the air, increases or decreases.

    If P = the ventilating pressure,
       L = the length of a drift,
       O = the perimeter, do.
Then LO = the rubbing surface,
And P varies as LO.

(2) The pressure per unit of area required to overcome

---

* P also varies as the density of the air; but the variations in density are so small that this may be neglected.

the friction of the air increases and decreases inversely as the sectional area of the drift increases and decreases.

If P = the ventilating pressure per unit of sectional area of the drift,
A = the sectional area of the drift,

Then P varies as $\dfrac{1}{A}$

(3) The pressure required to overcome the friction of the air increases and decreases in the same proportion that the velocity squared of the air increases and decreases.

If P = the ventilating pressure,
V = the velocity of the air,
Then P varies as V^2.

It follows from this that if P = the ventilating pressure per unit of sectional area,

(152.) $P = \dfrac{KLOV^2}{A}$

P may be obtained by means of a very delicate barometer, or by means of a water-gauge; L, O, A, and V, by measurement, treating each section of the mine separately. K is a constant to be found by experiment: it depends upon the units used. It is called the coefficient of friction; and is equal to the ventilating pressure required to overcome the resistance that a unit of air flying with unit velocity would meet with in circulating round a mine of unit area, and having unit rubbing surface.

When P is taken in lbs. per square foot.
L   do.   feet.
O   do.   feet.
V   do.   feet per minute.
A   do.   in square feet.

According to the experiments of MM. Devillez, Raux, &c., K = 0·000,000,009,36 (approximately) for the whole of a mine; but, in the case of clear smooth shafts alone, K = 0·000,000,003,6.

That is to say a ventilating pressure of 0·000,000,009,36 lbs.

per sq. ft. would be required to force 1 c. f. per min. through a mine 1 sq. ft. in area, and having a rubbing surface of 1 sq ft.

It is more convenient to take V in thousands of feet per min.; in which case K = 0·009,36 and 0·003,6 respectively.

A great many important facts may be deduced from formula (152), as see Fairley's "Ventilation of Coal Mines."

The Equivalent Orifice.—Air in passing through an opening in a thin plate meets with resistance; and M. Murgue has pointed out the convenience of assimilating the workings of a mine to such an opening in calculations for ventilating purposes. This opening he has named *the equivalent orifice.*

To find the equivalent orifice for any given mine :—

Let Q = Quantity of air in cubic feet per second passing through the opening (*i.e.*, circulating round the mine).

h_a = Ventilating pressure in feet of air column, required to overcome the resistance of the mine.

A = opening in thin plate in square feet (*i.e.*, equivalent orifice).

k = coefficient of contraction of orifice (*i.e.*, Vena contracta) = ·65.

Then—

$$Q = \sqrt{2gh_a} \times kA = \sqrt{2gh_a} \times ·65 A.$$

$$(153.) \therefore A = \frac{Q}{·65 \sqrt{2gh_a}}$$

But it is often more convenient to use the following units, viz. :—

Q in thousands of cubic feet per minute.
h_a in inches of water gauge.

When the formula becomes—

$$(154.) \quad A = \frac{0·37 Q}{\sqrt{h_a}} \therefore Q = \frac{A \sqrt{h_a}}{0·37} \text{ and } h_a = 0·1369 \left(\frac{Q}{A}\right)^2$$

The average value of A for English mines is said to be about 20, and for Belgian 8·6 sq. feet.

In order to find the quantity of air passing through a

drift, we must first find its velocity, and this is not easy, because the velocity varies in different parts of the same section of a gallery. The mean velocity may be found by dividing the section into squares with thin strings, each square about one foot in area, taking the velocity in each, and averaging them. M. Aguillon points out that, as the ratio between the mean velocity and the velocity at any given point of the same section is constant, it is only necessary to find this ratio once for all for any one convenient point in the section, and, in future, to measure the velocity at that point only.

A Difference of Pressure may be produced by:—

(A.) Reducing the density of the air in the upcast shaft (depressive, exhaustive, or negative ventilation) by means of:—

    (i.) The natural heat of the mine.
    (ii.) Furnace.
    (iii.) Steam jet.
    (iv.) Exhaust-fans.
    (v.) Varying capacity machines.

(B.) Increasing the density of the air in the downcast shaft (compressive, blowing, or positive ventilation) by means of:—

    (i.) Air-pump.
    (ii.) Water-fall.
    (iii.) Wind-cowl.
    (iv.) Blowing-fans.
    (v.) Varying capacity machines.

The first, Depressive Ventilation, is that adopted in practice almost without exception, though it is not easy to say why, except that the downcast is usually the drawing shaft, and it would not be convenient to have a ventilating machine upon it.

(A.) Depressive Ventilation.

(i.) *The Heat of the Mine* gives rise to what is usually called *natural ventilation*, and is the only means used for setting up a current of air in most metal mines. For the temperature of the strata at any given depth, see p. 77. In making calculations of the artificial pressure required in any case, the pressure due to the heat of the mine must not be forgotten. The formulæ below for the furnace are equally applicable to natural ventilation.

(ii.) *The Furnace:*—

If M = motive column (ventilating pressure) in feet of air of the temperature of the air in the upcast.

D = depth of upcast in feet (upcast column of heated air.)

t, T, t' = temperature of air in downcast, upcast, and returns respectively.

P = ventilating pressure in lbs. per square foot.

I = height of barometer in inches.

W = weight of one cubic foot of the return air in lbs.

Q = cubic feet of air per minute in the main return.

X = lbs. of coal burnt per hour.

Y = area of furnace in square feet.

Then—

(155.) $M = D \times \dfrac{T-t}{459+t}$

(156.) $T = \left\{ \dfrac{1\cdot 3253 \times I \times D}{\dfrac{1\cdot 3253 \times I \times D}{459+t} - P} \right\} - 459.$

(157.) $X = \left\{ \dfrac{WQ \times (T-t') \times 0\cdot 238 \times 60}{14,000} \right\} \times \text{say } 2.$

(158.) $Y = \dfrac{X}{10}$ (roughly.)

(159.) $P = \left(\dfrac{1\cdot 3253 \times I}{459+t} - \dfrac{1\cdot 3253 \times I}{459+T.} \right) \times D.$

No more air should pass through the furnace than is required for combustion, viz.:—About 300 cub. ft. per lb. of coal burned. In theory only about 150 cub. ft. are required.

The furnace should be placed as low down in the mine as possible, and should be built with coolers to prevent ignition of the seam.

As the heating of the air in the upcast dilates it; if the air be heated above a certain temperature its volume will become so great, and consequently its velocity in the upcast so high, that the resistance due to friction in the upcast will more than balance the increased ventilating pressure due to the increased temperature; and when this point is reached, the *more* the air in the upcast is heated the *less* will be the air circulated through the workings. Peclet considers that when the air in the upcast is expanded to twice its original volume the limit of furnace ventilation is reached.

The volume of air may be increased by building a cupola on to the upcast shaft.

If R = ratio that the volume of air required bears to the volume circulating.

D = depth of upcast in feet.
H = height in feet of cupola.

(160.) $H = (R^2 - 1) D$.

TABLE XXXI.

CONSUMPTION OF FUEL ON PIT FURNACES.

	Depth.	Coals per H.P. utilised per hour, excluding power due to heat of mine.	Coals per H.P. utilised per hour, including power due to heat of mine.	Quantity of air per min.
	feet.	lbs.	lbs.	cubic feet.
Thornley 5/4 Seam	556	85·5	37·5	45,756
Thornley Hutton Seam	868	162·4	57·1	26,574
Walker	960	30·5	15·6	44,800
Castle Eden	1,038	29·1	28·3	42,326
South Hetton	1,212	27·2	15·5	132,895
Wearmouth	1,800	29·5	7·9	70,500
Average		60·7	27·6	

(Trans. N.E.I., vi.)

(iii.) *The steam jet :*—If placed near the top of the pit will produce a current by pushing up the air above it; and the *maximum* units of work that it can yield will be given by formula (51), viz. :—$U = PV$. But if this steam were used with expansion in an engine to drive a fan the units of work *actually got in practice* would be more than PV. The steam jet, therefore, placed near the top of the shaft, cannot compete with a fan.

If placed at the bottom of the shaft it would also heat the upcast column of air; but the volume passing up the shaft would be augmented by the volume of the steam used, *i.e.*, the velocity and friction would be increased; and there would probably be some condensation and consequent downpour of water tending to reverse the current.

In practice it is found that the steam jet does not give such good results as the furnace. On the other hand, it cannot fire gas, and might (like the water fall) be used as a temporary expedient.

(iv.) *Fans* may be either centrifugal fans; or screw fans, as The Pelzer (see Trans. N.E.I., xxxi., abs. p. 9); about these last I do not propose to say anything.

Centrifugal Force:—

If F = Centrifugal force in lbs.
   W = Weight in lbs.
   V = Velocity in feet per second.
   R = Radius in feet.
   g = Force of gravity, say 32.

(161.) $F = \dfrac{W}{g} \times \dfrac{V^2}{R}$.

Centrifugal Fans.—These are either made of a large diameter, to run at a low velocity, as the Guibal and Waddle; or of a small diameter, to run at a high velocity, as the Bowlker, Capell, and Schiele. But the fundamental principle is the same in all, the work done depending upon the speed of the periphery, or tangential velocity as it is usually called.

Let H = gross ventilating pressure (theoretical depression) expressed in feet of air column of the density of the flowing air.

   u = Tangential velocity of fan in feet per second.

Then, in a theoretically perfect fan :—

(162.) $H = \dfrac{u^2}{g}$

But we have already seen formula (151) that $H = \dfrac{V^2}{64}$ *i.e.*, $= \dfrac{V^2}{2g}$. We may say, therefore, that :—In a perfect fan the theoretical depression is double the height due to the tangential velocity.

If H = as above.
P = Ventilating pressure in lbs. per sq. ft.
WG = Water gauge in inches.
d = density of water = 1000.
d' = density of air = 1·2 approx. at ordinary pressure and temperature.

Then—

(163.) $P = 5\cdot 2 \; WG \;\therefore\; WG = \dfrac{P}{5\cdot 2}$

(164.) $H = \dfrac{d \; WG}{d' \times 12} = \dfrac{1,000 \, WG}{1\cdot 2 \times 12} \;\therefore\; WG = \dfrac{1\cdot 2 \times 12 H}{1,000}$

The following formulæ are taken from Mr. A. L. Steavenson's translation of M. Murgue's work, to which students are referred for details :—

Let H = the theoretical depression in feet of air column that a perfect fan would give if its eye were shut off from the mine and atmosphere.

h_n = effective depression in feet of air column ; *i.e.*, the ventilating pressure required to overcome the resistance of the workings = the water gauge in the fan drift.

h_o = useless depression in feet of air column ; *i.e.*, the ventilating pressure required to overcome the resistance the air meets with in passing through the fan. To obtain this the communication between the fan-eye and the workings must be closed, and the eye connected direct with the atmosphere.

K

Q = Quantity of air in cubic feet per second.
A = Equivalent orifice in sq. ft. (see p. 124).
O = Orifice of passage in sq. ft.; *i.e.*, the area of a hole in a thin plate which would offer the same resistance to the air that the fan offers.
k = coefficient of efficiency of fan (varying in the case of well designed fans from 0·5 to 0·8).
g = gravity, say 32·19.
u = Tangential speed of fan in feet per second.

Then—

(165.) $H = h_a + h_o$

(166.) $H = \dfrac{u^2}{g}$

(See 153.) $A = \dfrac{Q}{0·65 \sqrt{2gh_a}}$

(167.) $O = \dfrac{Q}{0·65 \sqrt{2gh_o}}$

(168.) $h_a = \dfrac{Q^2}{2g\,(0·65A)^2}$

(169.) $h_o = \dfrac{Q^2}{2g\,(0·65O)^2}$

(170.) $Q = 0·65A \sqrt{2gh_a}$

(171.) $Q = 0·65O \sqrt{2gh_o}$

(172.) $\dfrac{h_o}{h_a} = \dfrac{A^2}{O^2}$

(173.) $h_a = \dfrac{H}{1 + \dfrac{A^2}{O^2}}$

(174.) $Q = \dfrac{0·65A \sqrt{2gH}}{\sqrt{1 + \dfrac{A^2}{O^2}}}$

And for practical calculations :—

$$(175.) \quad h_a = \frac{ku^2}{g\left(1 + \dfrac{A^2}{O^2}\right)}$$

$$(176.) \quad Q = \frac{0\cdot 65\sqrt{2kAu}}{\sqrt{1 + \dfrac{A^2}{O^2}}}$$

I believe Nasmyth was the first to apply the fan to ventilating purposes. His machine was an open running fan with straight blades, and gave a very small efficiency. To Guibal belongs the greater part of the credit for our present, comparatively speaking, perfect machine.

If Q = total volume of air circulating in a mine.
    q = a portion of Q due to any cause, say a fan.
    v = the rest of Q due to some other cause, say a furnace.

$$(177.) \quad Q = \sqrt{q^2 + v^2}$$

(v.) *Varying Capacity Machines.**—See description of Lemielle's, Nixon's, Struvé's, Cooke's, and Roots' blowers in Trans. N.E.I., i., vi., xi., xvi., xviii., xix., xxx.

(B.) Compressive Ventilation.

This has been tried in Germany (see Trans. N.E.I. xxxiv. abs. p. 49), though not in England, so far as I am aware. Many persons consider that such a method of ventilation would be preferable to the exhaustive system. (See *Colliery Guardian*, 1880, 2d part.) The waterfall is convenient as a temporary expedient; and either the water-fall or wind-cowl may be used for the permanent ventilation of a small non-fiery mine.

---

* These machines are best suited to mines with small equivalent orifices (less than 20 sq. ft.). There are not many of them in use in England.

Relation between Volume, Pressure, &c.

The ventilating pressure varies as :—
The depth of the upcast shaft (furnace ventilation).
The difference of temperature between upcast and downcast.
The HP of the ventilating machine or furnace.
The quantity of coals burned.

The quantity of air circulating varies as :—
The revolutions of the fan.
The tangential velocity of the fan.
The $\sqrt[2]{}$ of the ventilating pressure.
The $\sqrt[2]{}$ of the depth of the upcast (furnace ventilation).
The $\sqrt[2]{}$ of the difference of temperature (nearly, as see p. 126).
The $\sqrt[3]{}$ of the HP of the ventilating machine.
The $\sqrt[3]{}$ of the coals burned.

Authorities :—Trans. N.E.I., i. and vi., on the Steam Jet; xviii. and xix., on the Furnace; xxvi. and xxxi., on the Fan; Books and Papers by Fairley and Atkinson already mentioned; Murgue's "Theory and Practice of Centrifugal Ventilating Machines," by A. L. Steavenson.

TABLE XXXII.

TABLE OF HYPERBOLIC LOGARITHMS (FORMULÆ 61, 85, ETC.).

Nos.	Logarithm	Nos.	Logarithm	Nos.	Logarithm	Nos.	Logarithm
1·01	·009	1·11	·104	1·21	·190	1·31	·270
1·02	·019	1·12	·113	1·22	·198	1·32	·277
1·03	·029	1·13	·122	1·23	·207	1·33	·285
1·04	·039	1·14	·131	1·24	·215	1·34	·292
1·05	·048	1·15	·139	1·25	·223	1·35	·300
1·06	·058	1·16	·148	1·26	·231	1·36	·307
1·07	·067	1·17	·157	1·27	·239	1·37	·314
1·08	·076	1·18	·165	1·28	·246	1·38	·322
1·09	·086	1·19	·173	1·29	·254	1·39	·329
1·10	·095	1·20	·182	1·30	·262	1·40	·336

TABLE XXXII.—*continued.*

Nos.	Logarithm	Nos.	Logarithm	Nos.	Logarithm	Nos.	Logarithm
1·41	·343	1·81	·593	2·21	·792	2·61	·959
1·42	·350	1·82	·598	2·22	·797	2·62	·963
1·43	·357	1·83	·604	2·23	·802	2·63	·966
1·44	·364	1·84	·609	2·24	·806	2·64	·970
1·45	·371	1·85	·615	2·25	·810	2·65	·974
1·46	·378	1·86	·620	2·26	·815	2·66	·978
1·47	·385	1·87	·625	2·27	·819	2·67*	·982
1·48	·392	1·88	·631	2·28	·824	2·68	·985
1·49	·398	1·89	·636	2·29	·828	2·69	·989
1·50	·405	1·90	·641	2·30	·832	2·70	·993
1·51	·412	1·91	·647	2·31	·837	2·71	·996
1·52	·418	1·92	·652	2·32	·841	2·72	1·000
1·53	·425	1·93	·657	2·33	·845	2·73	1·004
1·54	·431	1·94	·662	2·34	·850	2·74	1·007
1·55	·438	1·95	·667	2·35	·854	2·75	1·011
1·56	·444	1·96	·672	2·36	·858	2·76	1·015
1·57	·451	1·97	·678	2·37	·862	2·77	1·018
1·58	·457	1·98	·683	2·38	·867	2·78	1·022
1·59	·463	1·99	·688	2·39	·871	2·79	1·026
1·60	·470	2·00	·693	2·40	·875	2·80	1·029
1·61	·476	2·01	·698	2·41	·879	2·81	1·033
1·62	·482	2·02	·703	2·42	·883	2·82	1·036
1·63	·488	2·03	·708	2·43	·887	2·83	1·040
1·64	·494	2·04	·712	2·44	·891	2·84	1·043
1·65	·500	2·05	·717	2·45	·896	2·85	1·047
1·66	·506	2·06	·722	2·46	·900	2·86	1·050
1·67	·512	2·07	·727	2·47	·904	2·87	1·054
1·68	·518	2·08	·732	2·48	·908	2·88	1·057
1·69	·524	2·09	·737	2·49	·912	2·89	1·061
1·70	·530	2·10	·741	2·50	·916	2·90	1·064
1·71	·536	2·11	·746	2·51	·920	2·91	1·068
1·72	·542	2·12	·751	2·52	·924	2·92	1·071
1·73	·548	2·13	·756	2·53	·928	2·93	1·075
1·74	·553	2·14	·760	2·54	·932	2·94	1·078
1·75	·559	2·15	·765	2·55	·936	2·95	1·081
1·76	·565	2·16	·770	2·56	·940	2·96	1·085
1·77	·570	2·17	·774	2·57	·943	2·97	1·088
1·78	·576	2·18	·779	2·58	·947	2·98	1·091
1·79	·582	2·19	·783	2·59	·951	2·99	1·095
1·80	·587	2·20	·788	2·60	·955	3·00	1·098

HYPERBOLIC LOGARITHMS.

TABLE XXXII.—continued.

Nos.	Logarithm	Nos.	Logarithm	Nos.	Logarithm	Nos.	Logarithm
3·01	1·101	3·41	1·226	3·81	1·337	4·21	1·437
3·02	1·105	3·42	1·229	3·82	1·340	4·22	1·439
3·03	1·108	3·43	1·232	3·83	1·342	4·23	1·442
3·04	1·111	3·44	1·235	3·84	1·345	4·24	1·444
3·05	1·115	3·45	1·238	3·85	1·348	4·25	1·446
3·06	1·118	3·46	1·241	3·86	1·350	4·26	1·449
3·07	1·121	3·47	1·244	3·87	1·353	4·27	1·451
3·08	1·124	3·48	1·247	3·88	1·355	4·28	1·453
3·09	1·128	3·49	1·249	3·89	1·358	4·29	1·456
3·10	1·131	3·50	1·252	3·90	1·360	4·30	1·458
3·11	1·134	3·51	1·255	3·91	1·363	4·31	1·460
3·12	1·137	3·52	1·258	3·92	1·366	4·32	1·463
3·13	1·141	3·53	1·261	3·93	1·368	4·33	1·465
3·14	1·144	3·54	1·264	3·94	1·371	4·34	1·467
3·15	1·147	3·55	1·266	3·95	1·373	4·35	1·470
3·16	1·150	3·56	1·269	3·96	1·376	4·36	1·472
3·17	1·153	3·57	1·272	3·97	1·378	4·37	1·474
3·18	1·156	3·58	1·275	3·98	1·381	4·38	1·477
3·19	1·160	3·59	1·278	3·99	1·383	4·39	1·479
3·20	1·163	3·60	1·280	4·00	1·386	4·40	1·481
3·21	1·166	3·61	1·283	4·01	1·388	4·41	1·483
3·22	1·169	3·62	1·286	4·02	1·391	4·42	1·486
3·23	1·172	3·63	1·289	4·03	1·393	4·43	1·488
3·24	1·175	3·64	1·291	4·04	1·396	4·44	1·490
3·25	1·178	3·65	1·294	4·05	1·398	4·45	1·492
3·26	1·181	3·66	1·297	4·06	1·401	4·46	1·495
3·27	1·184	3·67	1·300	4·07	1·403	4·47	1·497
3·28	1·187	3·68	1·302	4·08	1·406	4·48	1·499
3·29	1·190	3·69	1·305	4·09	1·408	4·49	1·501
3·30	1·193	3·70	1·308	4·10	1·410	4·50	1·504
3·31	1·196	3·71	1·311	4·11	1·413	4·51	1·506
3·32	1·199	3·72	1·313	4·12	1·415	4·52	1·508
3·33	1·202	3·73	1·316	4·13	1·418	4·53	1·510
3·34	1·205	3·74	1·319	4·14	1·420	4·54	1·512
3·35	1·208	3·75	1·321	4·15	1·423	4·55	1·515
3·36	1·211	3·76	1·324	4·16	1·425	4·56	1·517
3·37	1·214	3·77	1·327	4·17	1·427	4·57	1·519
3·38	1·217	3·78	1·329	4·18	1·430	4·58	1·521
3·39	1·220	3·79	1·332	4·19	1·432	4·59	1·523
3·40	1·223	3·80	1·335	4·20	1·435	4·60	1·526

HYPERBOLIC LOGARITHMS.

TABLE XXXII.—continued.

Nos.	Logarithm	Nos.	Logarithm	Nos.	Logarithm	Nos.	Logarithm
4·61	1·528	5·01	1·611	5·41	1·688	5·81	1·759
4·62	1·530	5·02	1·613	5·42	1·690	5·82	1·761
4·63	1·532	5·03	1·615	5·43	1·691	5·83	1·763
4·64	1·534	5·04	1·617	5·44	1·693	5·84	1·764
4·65	1·536	5·05	1·619	5·45	1·695	5·85	1·766
4·66	1·539	5·06	1·621	5·46	1·697	5·86	1·768
4·67	1·541	5·07	1·623	5·47	1·699	5·87	1·769
4·68	1·543	5·08	1·625	5·48	1·701	5·88	1·771
4·69	1·545	5·09	1·627	5·49	1·702	5·89	1·773
4·70	1·547	5·10	1·629	5·50	1·704	5·90	1·774
4·71	1·549	5·11	1·631	5·51	1·706	5·91	1·776
4·72	1·551	5·12	1·633	5·52	1·708	5·92	1·778
4·73	1·553	5·13	1·635	5·53	1·710	5·93	1·780
4·74	1·556	5·14	1·637	5·54	1·711	5·94	1·781
4·75	1·558	5·15	1·638	5·55	1·713	5·95	1·783
4·76	1·560	5·16	1·640	5·56	1·715	5·96	1·785
4·77	1·562	5·17	1·642	5·57	1·717	5·97	1·786
4·78	1·564	5·18	1·644	5·58	1·719	5·98	1·788
4·79	1·566	5·19	1·646	5·59	1·720	5·99	1·790
4·80	1·568	5·20	1·648	5·60	1·722	6·00	1·791
4·81	1·570	5·21	1·650	5·61	1·724	6·01	1·793
4·82	1·572	5·22	1·652	5·62	1·726	6·02	1·795
4·83	1·574	5·23	1·654	5·63	1·728	6·03	1·796
4·84	1·576	5·24	1·656	5·64	1·729	6·04	1·798
4·85	1·578	5·25	1·658	5·65	1·731	6·05	1·800
4·86	1·581	5·26	1·660	5·66	1·733	6·06	1·801
4·87	1·583	5·27	1·662	5·67	1·735	6·07	1·803
4·88	1·585	5·28	1·663	5·68	1·736	6·08	1·805
4·89	1·587	5·29	1·665	5·69	1·738	6·09	1·806
4·90	1·589	5·30	1·667	5·70	1·740	6·10	1·808
4·91	1·591	5·31	1·669	5·71	1·742	6·11	1·809
4·92	1·593	5·32	1·671	5·72	1·743	6·12	1·811
4·93	1·595	5·33	1·673	5·73	1·745	6·13	1·813
4·94	1·597	5·34	1·675	5·74	1·747	6·14	1·814
4·95	1·599	5·35	1·677	5·75	1·749	6·15	1·816
4·96	1·601	5·36	1·678	5·76	1·750	6·16	1·818
4·97	1·603	5·37	1·680	5·77	1·752	6·17	1·819
4·98	1·605	5·38	1·682	5·78	1·754	6·18	1·821
4·99	1·607	5·39	1·684	5·79	1·756	6·19	1·822
5·00	1·609	5·40	1·686	5·80	1·757	6·20	1·824

TABLE XXXII.—*continued*.

Nos.	Logarithm	Nos.	Logarithm	Nos.	Logarithm	Nos.	Logarithm
6·21	1·826	6·61	1·888	7·01	1·947	7·41	2·002
6·22	1·827	6·62	1·890	7·02	1·948	7·42	2·004
6·23	1·829	6·63	1·891	7·03	1·950	7·43	2·005
6·24	1·830	6·64	1·893	7·04	1·951	7·44	2·006
6·25	1·832	6·65	1·894	7·05	1·953	7·45	2·008
6·26	1·834	6·66	1·896	7·06	1·954	7·46	2·009
6·27	1·835	6·67	1·897	7·07	1·955	7·47	2·010
6·28	1·837	6·68	1·899	7·08	1·957	7·48	2·012
6·29	1·838	6·69	1·900	7·09	1·958	7·49	2·013
6·30	1·840	6·70	1·902	7·10	1·960	7·50	2·014
6·31	1·842	6·71	1·903	7·11	1·961	7·51	2·016
6·32	1·843	6·72	1·905	7·12	1·962	7·52	2·017
6·33	1·845	6·73	1·906	7·13	1·964	7·53	2·018
6·34	1·846	6·74	1·908	7·14	1·965	7·54	2·020
6·35	1·848	6·75	1·909	7·15	1·967	7·55	2·021
6·36	1·850	6·76	1·911	7·16	1·968	7·56	2·022
6·37	1·851	6·77	1·912	7·17	1·969	7·57	2·024
6·38	1·853	6·78	1·913	7·18	1·971	7·58	2·025
6·39	1·854	6·79	1·915	7·19	1·972	7·59	2·026
6·40	1·856	6·80	1·916	7·20	1·974	7·60	2·028
6·41	1·857	6·81	1·918	7·21	1·975	7·61	2·029
6·42	1·859	6·82	1·919	7·22	1·976	7·62	2·030
6·43	1·860	6·83	1·921	7·23	1·978	7·63	2·032
6·44	1·862	6·84	1·922	7·24	1·979	7·64	2·033
6·45	1·864	6·85	1·924	7·25	1·981	7·65	2·034
6·46	1·865	6·86	1·925	7·26	1·982	7·66	2·036
6·47	1·867	6·87	1·927	7·27	1·983	7·67	2·037
6·48	1·868	6·88	1·928	7·28	1·985	7·68	2·038
6·49	1·870	6·89	1·930	7·29	1·986	7·69	2·039
6·50	1·871	6·90	1·931	7·30	1·987	7·70	2·041
6·51	1·873	6·91	1·932	7·31	1·989	7·71	2·042
6·52	1·874	6·92	1·934	7·32	1·990	7·72	2·043
6·53	1·876	6·93	1·935	7·33	1·991	7·73	2·045
6·54	1·877	6·94	1·937	7·34	1·993	7·74	2·046
6·55	1·879	6·95	1·938	7·35	1·994	7·75	2·047
6·56	1·880	6·96	1·940	7·36	1·996	7·76	2·048
6·57	1·882	6·97	1·941	7·37	1·997	7·77	2·050
6·58	1·884	6·98	1·943	7·38	1·998	7·78	2·051
6·59	1·885	6·99	1·944	7·39	2·000	7·79	2·052
6·60	1·887	7·00	1·945	7·40	2·001	7·80	2·054

TABLE XXXII.—*continued.*

Nos.	Logarithm	Nos.	Logarithm	Nos.	Logarithm	Nos.	Logarithm
7·81	2·055	7·86	2·061	7·91	2·068	7·96	2·074
7·82	2·056	7·87	2·063	7·92	2·069	7·97	2·075
7·83	2·057	7·88	2·064	7·93	2·070	7·98	2·076
7·84	2·059	7·89	2·065	7·94	2·071	7·99	2·078
7·85	2·060	7·90	2·066	7·95	2·073	8·00	2·079

Specific Gravity.

As a cubic foot of water weighs 1,000 ozs., the weight of any substance can be got by multiplying its specific gravity by 1,000.

TABLE XXXIII.

SPECIFIC GRAVITIES.

Metals.

Platinum (laminated)	22·0690	Brass (cast)	8·3958
Pure Gold (hammered)	19·3617	Steel (hard)	7·8163
Gold 22 carat (do.)	17·5894	Iron (cast)	7·2070
Mercury	13·5681	,, (wrought)	7·7880
Lead (cast)	11·3523	Tin (cast)	7·2914
Pure Silver (hammered)	10·5107	Zinc (cast)	7·1908
Copper (cast)	8·7880		

Stones and Earth.

Marble (white Italian)	2·638	Portland Stone	2·145
Slate (Westmoreland)	2·791	Coal (Newcastle)	1·270
Granite (Aberdeen)	2·625	Brick (Red)	2·168
Paving Stone	2·4158	Clay	1·919
Mill Stone	2·4835	Sand (River)	1·886
Grindstone	2·1429	Chalk (mean)	2·315

Woods (Dry), &c.

Elm	0·588	Oak (English)	0·934
Fir (Riga)	0·753	Teak (Indian)	0·657
Larch	0·522	Cork	0·240
Mahogany (Spanish)	0·800	Sea water	1·027

(Twisden's "Mechanics.")

The specific gravity of the gases is given in Table XVI., p. 104.

EXAMPLES OF THE USE OF THE FORMULÆ.

(1.) What is the breaking load of a 10-inch hemp rope?
By formula (1) we find that :—
$W = 0.25 \times 10 \times 10 = 25$.
Answer, 25 tons.

(2.) What size of round iron-wire winding rope would be required for a pit 100 fathoms deep, and with a full cage weighing 3 tons, taking 8 as factor of safety?
By formula (11) we have :—

$$C = \sqrt{\dfrac{3}{\dfrac{1.5}{8} - \dfrac{100}{1.2 \times 2240}}} = \sqrt{19.96} = 4.47.$$

Anzwer, a $4\frac{1}{2}$ inch rope.

(3.) What must be the dimensions of a round, taper, plough steel, wire rope for a pit 500 fathoms deep, and a working load of 7 tons, at the top, bottom, and 100 fathoms from the bottom?
By page 18 we find that :—
The safe working load is 13,440 lbs. per sq. inch of section. The area, therefore, of the rope at the bottom must be $\dfrac{7 \times 2,240}{13,440} = 1.166$ sq. inches.

The area at the top by formula (14) and Table VII. is :—
$A = 1.166 \times 1.4549 = 1.696$.

The area at 100 fathoms from the bottom by the same formula and Table is :—
$A = 1.166 \times 1.0778 = 1.256$.

Answers,

Area at top	. . . .	1·166 sq. in.
,, bottom	. . .	1·696 ,,
,, 100 fms. from bottom	.	1·256 ,,

In this way the area of a taper rope at any point of its length may be obtained.

(4.) What should be the thickness of the plates of a single riveted iron boiler, 5 ft. diameter, to stand a working pressure of 40 lbs. above the atmosphere, taking 8 as factor of safety?

By formula (25) we have :—

$$TH = \frac{40 \times 5 \times 12}{50,000} \times 8 = 0\cdot384 \text{ inches.}$$

Answer, ⅜ inches.

(5.) What should be the thickness of an oak spherical dam with external radius 15 feet to withstand a pressure of 50 fathoms of water.

By formula (43) we have :—

r = 15 feet = 180 inches.
T = 10,000.
p = 50 fathoms = 130 lbs. per sq. in.

Then—

$$K = 180 \left\{ 1 - \sqrt[3]{1 - \frac{15 \times 130}{10,000}} \right\}$$

$$= 180 \left\{ 1 - \sqrt[3]{1 - 0\cdot195} \right\} = 180 \left\{ 1 - \sqrt[3]{0\cdot805} \right\}$$

$$= 180 \left\{ 1 - 0\cdot93024 \right\} = 12\cdot55.$$

Answer, 12·55 inches.

(6.) If the resistance offered by a set of railway carriages be 10 lbs. per ton, how many tons weight could a horse drag?

By Table XIV. we see that a horse can overcome a resistance of 120 lbs. He could therefore draw $\frac{120}{10} = 12$.

Answer, 12 tons.

(7.) What force must be exerted to draw a tub weighing 16 cwt. up an incline rising 2 inches to the yard, the tub, wheels and axles being $12\frac{1}{2}$ and $1\frac{1}{2}$ inches in diameter respectively?

By formula (131) :—

$$\text{Resistance} = mW + \frac{WH}{L};$$

and by formula (127) :—

$$m = 0\cdot 0882 \times \frac{1\cdot 5}{12.5} = 0\cdot 0106;$$

and we have given :—
W = 16 cwt. = 1,792 lbs.
H = 2 inches.
L = 1 yard = 36 inches.

$$\therefore R = 0\cdot 0106 \times 1,792 + \frac{1,792 \times 2}{36} = 118\cdot 55.$$

Answer, $118\frac{1}{2}$ lbs.

We see by Table XIV., p. 38, that a horse will exert a force of 120 lbs. Therefore one horse would be required to draw the tub.

(8.) At what inclination must a self-acting incline 600 ft. in length be laid in order that it may run a set of 10 tubs in one minute—the weight of a full tub being 15 cwt., of an empty tub 6 cwt., of the rollers and sheave 700 lbs., and of the rope 200 lbs.?

First find the resistance of friction as explained in formula (129), and let it be 16·8 lbs. for a full tub, and 6·72 lbs. for an empty tub, in other words, 0·01 of the weight. The friction of the rollers and sheave may be estimated at about 0·03 of their weight.

USE OF THE FORMULÆ.

Then—
L = 600, H = ?
F = (10 × 15 × 112) = 16,800 lbs.
E = (10 × 6 × 112) = 6,720 lbs.
T = 60 seconds.
$g = 32$
R = 200 lbs.
S = 700 lbs.
$m = 0·01, m' = 0·03$.
W = (16,800 + 6,720 + 200 + 700) = 24,420 lbs., and by formula (138):—

$$\frac{H}{L} = \frac{·01(16,800 + 6,720) + ·03 \times 700 + \frac{24,420 \times 2 \times 600}{32 \times 60 \times 60}}{16,800 - (6,720 + 200)} =$$

$$\frac{235 + 21 + 254}{16,800 - 6,920} = \frac{510}{9,880}$$

$$\frac{H}{L} = \frac{510}{9,880} \therefore H = \frac{600 \times 510}{9,880} = 30·96.$$

That is to say, the height of the incline must be 30·96 ft. which, in a length of 600 ft., is equal to nearly 1 in 20, or 1·8 in. to the yard.

In laying out the incline, the average gradient should be 1·8 in. per yard; but it should be rather steeper at the top, and rather flatter at the bottom.

(9.) What will be the position of meetings in a pit 237½ fathoms deep when the rope is ·71 inches thick, the diameter of the drum at the lift 16 ft. 2 in., and the revolutions of the drum 25⅝ per winding?

By formula (115):—

$$n = \frac{25\frac{5}{8}}{2} = 12·8125.$$

r = 97·355 ∴ 2r = 194·71.
t = 0·71.
Then:—

$$d = 3·1416 \times 12·8125 \left\{ 194·71 + (11·8125 \times ·71) \right\} =$$

40·2517 × 203·097 = 8,174 inches, and 8,174 inches is 113·51 fathoms.

Answer, 113½ fathoms from the bottom of the shaft.

(10.) Design a non-condensing pumping engine to force 1,000 gallons per minute, 50 fathoms vertically, with a pressure of 30 lbs. of steam above the atmosphere.

The pressure in the boiler will be $30 + 14\cdot7 = 45$ lbs. (say), and the steam, therefore, may be cut off at $\frac{2}{5}$ of the stroke, reducing the pressure of the exhaust to $\frac{2}{5} \times 45 = 18$ lbs., or 3 lbs. above the atmosphere.

The mean pressure by formula (61) will be :—

$$\frac{45 \times 2}{5}\left(1 + \text{Hyp. log. } \frac{5}{2}\right) = 18\,(1 + \text{Hyp. log. } 2\cdot5).$$

By Table XXXII., p. 132, Hyp. log. $2\cdot5$ is $0\cdot916$ and $18 \times 1\cdot916 = 34\cdot488$ lbs.

Deduct from this the back pressure, *i.e.*, the pressure of the atmosphere, and we get :—

Mean effective pressure on piston = 20 lbs. (say).

The work to be done, expressed in units of horse-power, is by formula (49) :—

$$\frac{10 \times 1,000 \times 50 \times 6}{1 \times 33,000} = 90\cdot9.$$

Add 50 °/$_\circ$ for friction and we get (say) 136 as the horse-power of the engine required.

We will assume 6 feet as the length of stroke. (If on trial it is not suitable, we must assume some other length, and make our calculations over again.) Then by formula (122) the speed of piston should be :—

$\sqrt{6} \times 60 = 2\cdot45 \times 60 = 147$ ft. per minute, and the area of the piston by formula (60) is :—

$$\frac{33,000 \times 136}{20 \times 147} = 1,526 \text{ sq. inches;}$$

and the diameter by formula (65) is :—

$$\sqrt{\frac{1,526}{\cdot7854}} = 44\cdot1 \text{ inches (say) 45 inches.}$$

The ram for pumping the water should be double acting, and will have the same stroke as the engine, viz. :—6 ft.

Allowing 5°/₀ loss of water due to leakage, etc., its diameter by formula (121) will be :—

$$\sqrt{\frac{1{,}050}{0{\cdot}034 \times 6 \times 24{\cdot}5}} = \sqrt{210} = \text{(say) a } 14\tfrac{1}{2}'' \text{ ram.}$$

The answer then is :—
An engine with one 45" cylinder, by 6' stroke; and a double-acting ram $14\tfrac{1}{2}''$ diameter, by 6' stroke.

(11.) How many egg-ended boilers 36' long by 5' diameter will be required to drive the above engine?

Each boiler will have a fire-grate area of 25 sq. ft. And the effective heating surface in sq. yards (see p. 49) will be :—

$$\frac{3{\cdot}1416 \times 5 \times 36}{9} \times \frac{3}{8} = 23{\cdot}56 \text{ sq. yards.}$$

The number of cubic ft. of water evaporated into steam per hour will be by formula (75) :—

$$\sqrt{23{\cdot}56 \times 25} = 24{\cdot}27 \text{ cubic ft. per hour};$$

and this is equal, by Table XVI., to $24{\cdot}27 \times 562 = 13{,}640$ cubic ft. of steam at 45 lbs.

The engine consumes :—

$$\frac{2}{5} \times \frac{1526}{12 \times 12} \times 6 \times 24{\cdot}5 \times 60 = 37{,}387 \text{ cubic ft. of steam per}$$
hour, and will, therefore, require :—

$$\frac{37{,}387}{13{,}640} = 2{\cdot}74 \text{ boilers, say 3 boilers.}$$

(12.) How many tons of small coal would the above boilers consume per fortnight?

As the engine does not require the full power of the boilers, we may take the consumption of coal at 18 lbs. per sq. foot of grate per hour, instead of at 20 (see p. 49); and the consumption will be :—

$$\frac{18 \times 25 \times 3 \times 24 \times 14}{2{,}240} = 200 \text{ tons (say).}$$

(13.) What should be the dimensions of a chimney to supply draught for two Lancashire boilers, each having 80 ft. of flue length, and each consuming 300 lbs. of coal per hour?

By p. 126, 300 lbs. of coal requires 90,000 cubic ft. of air to burn it, and this air will be doubled in volume when discharged from the chimney. For the two boilers, therefore, 360,000 cubic ft. of air per hour must be discharged, which is equal to 100 cubic ft. per second. We may take 20 ft. per second (in practice the velocity of discharge varies very much) as a convenient velocity of discharge, from which it follows that the area of chimney at the top should be 5 sq. ft., say $2'.7''$ in diameter.

Then by formula (78) we have:—

L = height of chimney, including length of flue (where there is more than one boiler we do not take the aggregate length of the flues, but the length of one only). For the height of the chimney we must assume a quantity, say 50 ft., so that L = 130, and v = 20, and D = $2'.7''$, say 2·6 ft.

Then—

$$h = \frac{20 \times 20}{2 \times 32 \cdot 2} \times \left(13 + \frac{0 \cdot 048 \times 130}{2 \cdot 6}\right) = 95 \cdot 48 \text{ ft.}$$

To produce this motive column we require by formula (79) a chimney:—

$$H = \frac{95 \cdot 48}{\left(\cdot 96 \times \frac{1059}{519} - 1\right)} = \frac{95 \cdot 48}{0 \cdot 9584} = 100 \text{ ft.}$$

We see then that our assumption of 50 ft. for the height of the chimney was too little, and that with a velocity of 20 ft. per second we should require a chimney tall and narrow. We will, therefore, try a height of 64 ft., and a velocity of 16 ft. per second. The area at the top, for a discharge of 16 ft. per second, is $\frac{100}{16} = 6 \cdot 25$ sq. ft., which corresponds to a diameter of 2·82 ft. And:—

$$h = \frac{16 \times 16}{2 \times 32}\left(13 + \frac{\cdot 048 \times 144}{2 \cdot 82}\right) = 4 \, (15 \cdot 45) = 61 \cdot 80 \text{ ft., and}$$

by formula (79):—

$$H = \frac{61 \cdot 8}{\left(\cdot 96 \times \frac{1059}{519} - 1\right)} = 63 \cdot 4$$

64 ft. then is rather too high, and we may say that a chimney 62 ft. high, and 2′.10″ diameter at the top, would satisfy our requirements.

The inside diameter at the bottom may be the same as the inside diameter at the top; but should not be less, and is usually made rather more.

(14.) How many units of work are required to compress 1 lb. of air at 68°, and under a pressure of one atmosphere to $\frac{1}{6}$ of its volume: 1st, isothermally; 2nd, adiabatically?

(1.) Isothermally, by formula (85).

$P_1 = 14·7 \times 144 = 2,116·8$ lbs. per sq. ft.
$V_1 = 13·3$ cubic feet.
$\frac{V_1}{V_2} = 6$; the hyp. log. of which by Table XXXII., is 1·79.

Then:—
$U = 2,116·8 \times 13·3 \times 1·79 = 50,394.$

(2.) Adiabatically, by formulæ (87 & 84).

$T_1 = 68 + 459 = 527$ absolute temp.
$W =$ one lb. *i.e.* $= 1.$

$T_2 = $ (by formula 84) $527 \left(\frac{6}{1}\right)^{·408}$

Log. $6 = 0·7781513$
           ·408

62252104
311260520

·3174857304 corresponds to 2·077.

∴ $T_2 = 527 \times 2·077 = 1,094.$

Then, by formula (87):—
$U = 130·3 \, (1,094 - 527) = 73,680.$

Answers:—
1st. Isothermally, 50,394 units of work.
2nd. Adiabatically, 73,680   do.

(15.) A, B, and C are three bore holes; the depths of which, from the same horizontal plane to a seam of coal, are respectively 100, 106, and 108 yards. From A to B is 100 yards, and from A to C 120 yards. The angle in a horizontal plane between A B and A C is 30°. What is the direction of the dip of the seam, and the angle of dip?

By formulæ (105 & 106).
$a = 100$; $a' = 120$; $W = 30$;
$d = 8$; $d' = 8$.

Then:—

$$\text{Tan } V = \frac{\frac{6 \times 120}{8} \times 0\cdot 5}{100 - \left(\frac{6 \times 120}{8} \times 0\cdot 866\right)} = 2\cdot 04 = 63° \ 53'.$$

$$\text{Tan } S = \frac{8}{100 \times 0\cdot 8979} = 0\cdot 089 = 5° \ 7'.$$

The dip is at right angles to the strike, and the strike of the bed makes, we see, an angle of 63° 53' with the line A B. And the angle of dip is 5° 7'.

Suppose B to be due north of A, and C to lie on the east side of B: the seam will dip 5° 7'; N. 26° 7' E.

(16.) What is the cost of boring a hole on the diamond system, 250 fathoms deep? 250 fathoms is 1,500 feet, and by p. 72, the price for the first step is £30, the number of steps is 15, and the increase in price for each step is £30.

Then by formula (107):—

$$c = \left\{ 2 \times 30 + (15 - 1) \ 30 \right\} \frac{15}{2} = 3,600.$$

Answer £3,600.

(17.) What is the ventilating pressure in lbs. per square foot required to circulate 10,000 cubic feet per minute through a drift two miles long:—

1st. When the drift is circular, 7·98 feet in diameter.
2nd. When the drift is square, 7·071 feet high.
3rd. When the drift is oblong, 5 feet × 10 feet?

We first note that the area of each of these drifts is practically the same, viz., 50 square feet; and that, there-

fore, the velocity of the air will be the same in each, viz. :—
$\frac{10,000}{50}$ = 200 feet per minute; or 0·2 thousands of feet per minute. But the perimeters vary.

By formula (152).

$$P = \frac{K L O V}{A} = \frac{K L V^2}{A} \times O.$$

$$= \frac{·009 \times 2 \times 1760 \times 3 \times 0·2 \times 0·2}{A} \times O.$$

$$= 0·076 \times O.$$

Then—

   1st. (circular drift) P = 0·076 × 25·07 = 1·90.
   2nd. (square drift) P = 0·076 × 28·28 = 2·15.
   3rd. (oblong drift) P = 0·076 × 30·00 = 2·28.

Answer.—1·90, 2·15, and 2·28 lbs. per square foot respectively. From which we see that the circular drift offers the least resistance.

(18.) If the depth of the shafts of a mine ventilated by a furnace be 1,000 feet, the temperature of the downcast 41°, of the upcast, 141°, and the height of the barometer, 30 inches; what will the ventilating pressure be in lbs. per square foot?

By formula (159) :—

$$P = \left(\frac{1·3253 \times 30}{459 + 41} - \frac{1·3253 \times 30}{459 + 141} \right) \times 1,000.$$
$$= (0·079518 - 0·066265) \times 1,000.$$
$$= 13·253.$$

Answer.—13¼ lbs. per square foot.

(19.) What should be the indicated horse-power of a hauling engine to work a plane 1,200 yards in length with two curves of 58° and 82° respectively: the maximum work being to draw the full set, weighing 35,840 lbs., up a bank at the inbye end, rising 1 in 12, at the rate of 4 miles an hour? Assuming that the coefficient of friction of the tubs = ·01; the weight of ropes, rollers, and sheaves =

21,000 lbs.; the main rope drums at curves are 18" diameter with $3\frac{1}{2}$" axles—

Then the pull, or force, to be exerted by the engine is:—

1. Friction of set 35,840 × ·01, as see formula (127), = ... 358·40
2. Gravity of set 35,840 ÷ 12, as see formula (130), = ... 2,986·66
3. Friction of rope on rollers, &c., 21,000 × ·03, as see m', p. 98, = ... 630·00

  Total resistance on straight road ...... 3,975·06

To this must be added the resistance due to the friction of the main rope upon the drums at the curves, viz:—

First curve with angle of 58°, the pressure upon the drums (or sheaves, as the case may be) by the parallelogram of forces (see Twisden's "Practical Mechanics," 4th edition, pp. 54 and 83), will be:—

$$\text{Pressure} = \frac{3,975 \sin 58}{\sin 151} = \frac{3975 \times \cdot 848}{\cdot 484} = 6,964 \cdot 4$$

Second curve with angle of 82° gives in the same way:—

$$\text{Pressure} = \frac{3,975 \sin 82}{\sin 139} = \frac{3,975 \times \cdot 99}{\cdot 656} = 5,998 \cdot 8$$

The total pressure on the drums at the curves is therefore 12,963 lbs.; and the friction due to this pressure is, by formula (114)—

$$\frac{12,963 \times \cdot 07 \times 3 \cdot 5}{18} = 176 \cdot 44 \text{ lbs.}$$

The total force the engine must exert is therefore:—

1. Friction of set .... 358·40 lbs.
2. Gravity of set .... 2,986·66 „
3. Friction of rope .... 630·00 „
4. Friction of rope at curves ...... 176·44 „

  Total................ 4,151·50 lbs.

The speed is 4 miles an hour, which is 352 feet per minute. The horse-power therefore by formula (49) is:—

$$\frac{4{,}151\cdot5 \times 352}{1 \times 33000} = 44\cdot28$$

Allowing an efficiency of 50%, the hauling engine would have to be of about 90 horse-power.

Answer.—90 indicated horse-power.

The conditions would be fulfilled by a non-condensing engine with two cylinders 14" diameter × 2' 4" stroke, running at 75 revolutions, with 30 lbs. of steam, geared 3 to 1 to two drums, 4' 6" diameter by 2' wide.

(20.) It is proposed to increase the quantity of air circulating round a mine, 10,000 cubic feet per minute, by building a chimney (or cupola, as it is called in the North of England) on to the upcast shaft. The original volume is 60,000 cubic feet, and the upcast shaft is 400 feet in depth. What must be the height of the cupola?

The ratio the required volume bears to the volume circulating is $\frac{70{,}000}{60{,}000} = \frac{7}{6}$. Therefore by formula (160) we have:—

$$H = \left(\frac{7 \times 7}{6 \times 6} - 1\right) \times 400 = 0\cdot36 \times 400 = 144.$$

Answer.—144 feet.

(21.) What size of pipe would be required to supply an engine to be placed 1,200 yards from the boilers with 2 cubic feet per second of steam at a pressure of 30 lbs. above the atmosphere, the pressure in the boilers being 45 lbs. above the atmosphere?

First approximation:—Let us assume that one cubic foot per second will be lost by condensation; then, in order to deliver two cubic feet at the engine, the boilers must supply three cubic feet; and the mean volume squared passing through the pipe will be, by formula (100),

$$Q^2 = \frac{(3 \times 3) + (3 \times 2) + (2 \times 2)}{3} = 6\cdot333;$$

and the size of pipe to pass this quantity of steam, with the loss of pressure we can afford, viz., 15 lbs., will be got from formula (99) as follows:—

$$a = \frac{1,000,000 \ (45-30)}{1,200 \times 6\cdot 333 \times 2\cdot 06} = 958;$$

and by Table XXI. this value of a corresponds with a pipe very nearly 3 inches in diameter.

We must now find the loss by condensation with a 3 inch pipe. Its external diameter would be $3\frac{1}{2}$ inches, and if coated with $1\frac{1}{2}$ inches of non-conducting material, the total outside diameter would be $6\frac{1}{2}$ inches; and the surface in square feet of a $6\frac{1}{2}$-inch pipe, 1,200 yards long, is:—

$$\frac{6\cdot 5 \times 3\cdot 1416}{12} \times 1,200 \times 3 = 6,126 \text{ square ft.}$$

The differences in temperature between the surface of the pipe, and the air in the drift, and the drift sides, will depend upon the non-conducting composition used, the depth of the pit, and the nature of the drift. We will assume that the drift is rather a small dry return without much air passing, and that the pit is a shallow one.

In this case we might suppose that:—

Temperature of pipe surface = 120°
    do.   air in drift = 82°
    do.   drift sides = 80°

Then by formulæ (89 & 92) we have—

$U = \cdot 74 \times 40 \times 6,126 \times 1\cdot 28 = 232,102$
$U_1 = 38 \times 1\cdot 14 \times \cdot 5154 \times 6,126 = 136,775$

Total units of heat lost . $= \underline{368,877}$

and the mean pressure of the steam being $37\frac{1}{2}$ lbs. above the atmosphere, we have by formula (90) and Table XVI.:

$$L = \frac{368,877}{915} = 403\cdot 14 \text{ lbs. of steam condensed per hour.}$$

But one cubic ft. of steam, at a total pressure of $52\frac{1}{2}$ lbs., weighs, by Table XVI., 2·05 oz. = 0·128 lbs.; so that the number of cubic feet of steam condensed per second will be:—

$$\frac{403\cdot 14}{\cdot 128 \times 60 \times 60} = 0\cdot 875.$$

We will now calculate the loss of pressure.

By formula (100) the mean volume squared is:—

$$Q^2 = \frac{(2\cdot 875 \times 2\cdot 875) + (2\cdot 875 \times 2) + (2 \times 2)}{3} = 6;$$

and the loss of pressure by formula (101) is:—

$$P - p = \frac{1{,}200 \times 812 \times 2\cdot 06 \times 6}{1{,}000{,}000} = 12\cdot 04.$$

Say 12 lbs. But we are at liberty to lose 15 lbs., therefore a 3-inch pipe is a little too large.

We might make a second approximation and so calculate the exact size theoretically; but the size would be so nearly 3 inches, that for practical purposes this size would be required.

INDEX.

ACCIDENTS in mines, 116
   Acid water corrodes pumps, 91
Acre of coal, 14
   of water, 39
Act, explosives, 28
Adits, 78
Air and gases, general properties of, 101
   compressed, 54
   compressed, friction of, 56
   friction of, in workings, 122
   quantity of, circulating, 121
   temperature of, in upcast, 126
   vessel, 91
   weight of, 102
Ambulance classes, 116
Amount and direction of dip from three bore-holes, 71
Analyses of air, 120
   of fire-damp, 105
Area of boiler-grate, 49
   of furnace fire-grate, 126
Asphyxia, 117
Atmosphere, pressure of, 101
Atomic weights, 107

BAROMETER, heights measured by, 102
Battens, 26
Beams, strength of, 21
   circular, strength of, 22
Boilers, 45
   consumption of fuel, 49
   corrosion, 48
Boiler doctors, 48
   explosions, 47
   fittings, 46
   horse-power of, 49

Boiler, incrustations of, 47
   strength of, 20, 21, 45
   surface of, 49
   weight of, 46
Bord and pillar system of working, 80
Boreholes, amount and direction of dip of seam from, 71
   cost of, 72
   water delivered by, 93
Boring against old workings, 93
Brattice, 76
Brick pillars, 21

CAGES, 87
   Carbonic acid gas, 112
Carbonic oxide, 109
Cast-iron girders, strength of, 23
   pipes, strength of, 20
   pipes, weight of, 20
   tubbing, 25
Centrifugal fans, 128
   force, 128
Certificated managers' examination, 4
Chains, strength of, 20
Chaudron's tubbing, 75
Chavette method of sinking, 75
Chemical symbols, 106
Chemistry, 2, 106
Chimneys, 52
Circles, dimensions of, 17
Coal burnt in flue boiler, 49
   burnt on ventilating furnace, 127
   produce of, 119
Coalfields, 6, 11, 14
Coefficient of friction on level rolley-way, 95
Coefficient of friction, rollers and sheave, 98

Coefficient of friction, ventilation, 123
Coefficient of rupture, 22
Column of water, pressure of, 92
Columns, long square, 23
Composition of coal, 14
Compound engine, area of small cylinder, 43
Compounds and elements, 106
Compressed air, 54
Compressive ventilation, 131
Condensation, 52
Condensing engine, 42
Conical drum counterbalance, 86
Consumption of coal in boilers, 49
    of provender, 36
    of powder, 30
    of timber, 27
Contact with air, loss by, 59
Corrosion of boilers, 48
    of pumps by acid water, 91
Cost of boring, 72
    haulage, 99
    horses, 36
    pumping, 93
    sinking, 76
    timbering, 27
    walling, 76
    working, 84
Costeaning, 71
Counterbalances, 85
Course of study, 1
Cupola, ventilation by, 127
Curves, elevation of outer rail, 97
    resistance on, 96
Cylinders, measurement of, 44
Cylindrical dam, walling or tubbing, 24

DAMS, tubbing and walling, 24, 94
Deals, 26
Deep boreholes, 72
Deep mines, 77
Depressive ventilation, 125
Diamond process, 72
Dimensions of circles, 17
    cylinders, 44
    hanging on place, 77
Direction and amount of dip from three boreholes, 71

Discharge of water from old workings, 94
Draining, 89
Drawing, mechanical, 3
Drawings of coal, 13
Dykes, 12

EFFECTIVE horse-power, 43
    of steam-engine, 34
Electric current, horse-power of, 67
Electricity, 67
Elements and compounds, 106
Elevation of outer rail on curves, 97
Endless chain or rope, 99
Energy, 33
Engines, horse-power of, 42
Equations, chemical, 108
Equivalent orifice, 124
Explosions, 47, 113
Explosives, 28, 30

FALLING body, velocity of (same as for wind), 122
Fall of rain per in. per acre, 39
Fans, 128
Feed water, 48
Feeders, measurement of, 92
Fire-damp, 112
    analyses, 105
    detectors, 70, 115
    explosion of, 113
    method of dealing with, 115
Fire-grate, area in boilers, 49
    area in ventilating furnaces, 126
Flat-ropes, 19
Fluid, work furnished by discharge of, 39
Food, horse's, 35
Force, centrifugal, 128
Formulæ, chemical, 107
    examples of the use of, 138
Friction of air, coefficient of, 123
    of air round workings, 122
    compressed air, 56
    pulleys, 87
    rolley-ways, 95
    water in pipes, 92
    loss by, in steam transmission, 62

INDEX.

Fuel consumed in boilers, 49
    furnaces, 127
Furnace, coals burnt on, 127
    ventilating pressure produced by, 126

GALLONS of water delivered by borehole, 93
    pump, 91, 92
Gas, diffusion of, 104
    illuminating, 118
    in goaves, 102
    pressure of, in coal, 103
Gases, the, 109
    occluded, 103
    specific gravity of, 104
    transpiration of, 103
    weight of, 102
Geology, 3
Girders, strength of, 23
Gradient at which traction in-bye and out-bye is the same, 96
Grate of boiler, area of, 49
    of ventilating furnace, area of, 126
Guides, 76
Gunpowder, composition of, 30

HAND-boring, 72
    Hanging-on place, dimensions of, 77
Haulage, underground, 95
Heat, units of, 33, 34
Heights, by barometer, 102
Horses, 35, 90, 97
Horse-power, effective, 43
    nominal, 43
    of boilers, 49
    electric current, 67
    engines, 42
    waterfalls, 38
    windmills, 40
    units of, 33
Hydraulic motors, 38
Hydrogen, 110
Hydrogen sulphide, 111
Hyperbolic logarithms, table of, 132

ILLUMINATING gas, 118
    Inclination of roads, resistance due to, 96
Incline counterbalance, 85
Inclines, self-acting, 97
Incrustation of boilers, 47
Inertia, 85
Iron pipes, cast, strength of, 20
    weight of, 20

KEEPS, 76
    Kind-Chaudron method of sinking, 74
Koepe-system of winding, 86

LANGUAGES, modern, 3
    Lighting, 69
Lime cartridge, 31
Load of timber, 26
Logarithms, hyperbolic, table of, 132
Long square columns, strength of, 23
Long wall, 80

MACHINERY, 32
    Magnetism and electricity, 67
Main and tail-rope, 99
Main-rope planes, 99
Marsh-gas, 112
    explosion of, 113
Masonry pillars, 21
Mathematics, 2
Measure of timber, 26
Meetings, to find, 88
Men and horses, 35
Mineral products, 12
Mines Act, 74
Molecules, 107
Motive column in feet of air (fan), 129
    (furnace), 126
    (water-gauge), 129
Motors, 97

NITROGEN, 111
    Nominal horse-power, 43

OCCLUDED gases, 103
    Old workings, borings against, 93
    emptying water from, 94
Ordinary method of sinking, 74

Orifice, equivalent, 124
   of passage, 130
Overhand stoping, 83
Oxygen, 109

PASSAGE, orifice of, 130
   Pendulum counterbalance, 86
Physics, 2
Pillars, brick, strength of, 21
Pipes, cast-iron, strength of, 20
   cast-iron, weight of, 20
   friction of air in, 56
   friction of water in, 92
   friction of steam in horizontal, 60
   friction of steam in vertical, 61
Planks, 26
Poetsch's method of sinking, 74
Power, transmission of, 54, 70
   units of, 33
Pressure of steam, 50
   wind, 40
Prime movers, 35
Produce of coal (coke, gas, &c.), 119
   coal-seams, 14
Properties of air and gases, 101
   steam, 50
Proportions of engines, 44
Props, 26
Provender, 35
Pulleys, 87
Pulley-frame legs, strength of, 23
Pump corroded by acid water, 91
   stroke of, 92
   water delivered by, 91, 92
Pumping by men and horses, 90

QUANTITY of air circulating, 121

RADIATION, loss by, in steam transmission, 58
Rainfall per inch per acre, 39
Rappers, 76
Red water, 91
Relation between volume, pressure and air circulating, 132
Resistance on curves, 96
   to traction on underground rolley-ways, 95
Respiration, artificial, 117

Ropes, 16, 19, 57
Round ropes, 16, 20
   taper-ropes, 18

SAFETY-LAMPS, 116
   Safety-valves, 46
Sea water, specific gravity of, 137
Sedimentary rocks, 7
Search for minerals, 71
Self-acting inclines, 97
Shaft fittings, 75
   gates, 76
   pillars, 76
Shot-firing, 68
Signalling, 69
Sinking, ordinary method of, 74
Specific gravity of gases, 104
   solids, 137
Spherical dam, 24
Square of flooring, 26
Standard of timber, 27
Statistics, 9
Steam engine, 40
Steam jet, 127
Steam, pressure, &c., of, 50
   transmission of, 58
Stoping, 83
Strength of materials, 16
Surface of boiler, or any cylindrical vessel, 49
Sylvester's artificial respiration, 117
Symbols and atomic weights, 107
Syphon, 89

TAIL-ROPE COUNTER-BALANCE, 86
Taper-ropes, 18
Temperature, absolute, 41
   of compressed air, 55
   of upcast, 126
Timber, 26
Timbering, cost of, 27
Transmission of power, 70
Transpiration of gases, 103
Troubles, 12
Tubbing, 75, 94
   strength of, 25
Tubes, boiler, 20
Tubs, capacity of, 99
   friction of, 95

INDEX.

UNDERGROUND temperature, 77
Underhand stoping, 83
Units of heat, 33, 34
    horse-power, 33
    power, 33
    work from fluids' discharge, 39
    work stored in waterfall, 39
    work stored in wind, 40

VARYING capacity machines, 131
Velocity of the wind, 122
Ventilation, 122
    compressive, 131
    depressive, 125
Ventilating pressure, fan, 129
    in feet of air, 126, 129
    furnace, 126
    in lbs. per sq. ft., 123
    water-gauge, 129
Vitiation of air, 120

WALLING, 75
    Water delivered by boreholes, 93
    pipes, 92
    pump, 91, 92
Water gauge, 129

Water, measurement of feeders of, 92
    pressure of, 92
Water rings, 76
Water, friction of, in pipes, 92
    weights and measures of, 93
Wedges, 30
Weight of air and of any gas, 102
    boilers, 46
    coal seams, 15
    ores, 15
    of provender, 37
    pipes, 20
    ropes, 17
Weights, atomic, 107
    of various substances, 137
Wind, pressure of, 40
    units of work from, 40
    velocity of, 122
Winding engines, 85
Windmills, 39
Wire-ropes, 57
Work furnished by fluid's discharge, 39
    of horses, 38, 90, 97
    men, 37, 90
    units of, and horse-power, 33
Working, systems of, 79
Wormald's composition, 64
Wrought-iron girders, strength of, 23

THE END.

7, STATIONERS' HALL COURT, LONDON, E.C.
October, 1886.

A CATALOGUE OF BOOKS
INCLUDING MANY NEW AND STANDARD WORKS IN
ENGINEERING, ARCHITECTURE, MECHANICS,
MATHEMATICS, SCIENCE, THE INDUSTRIAL ARTS,
AGRICULTURE, LAND MANAGEMENT,
GARDENING, &c.
PUBLISHED BY
CROSBY LOCKWOOD & CO.

CIVIL ENGINEERING, SURVEYING, etc.

The Water Supply of Cities and Towns.
A COMPREHENSIVE TREATISE on the WATER-SUPPLY OF CITIES AND TOWNS. By WILLIAM HUMBER, A-M. Inst. C.E., and M. Inst. M.E., Author of "Cast and Wrought Iron Bridge Construction," &c., &c. Illustrated with 50 Double Plates, 1 Single Plate, Coloured Frontispiece, and upwards of 250 Woodcuts, and containing 400 pages of Text. Imp. 4to, £6 6s. elegantly and substantially half-bound in morocco.

List of Contents.

I. Historical Sketch of some of the means that have been adopted for the Supply of Water to Cities and Towns.—II. Water and the Foreign Matter usually associated with it.—III. Rainfall and Evaporation.—IV. Springs and the water-bearing formations of various districts.—V. Measurement and Estimation of the flow of Water —VI. On the Selection of the Source of Supply.—VII. Wells.—VIII. Reservoirs.—IX. The Purification of Water.—X. Pumps. — XI. Pumping Machinery. — XII. Conduits.—XIII. Distribution of Water.—XIV. Meters, Service Pipes, and House Fittings.—XV. The Law and Economy of Water Works. XVI. Constant and Intermittent Supply.—XVII. Description of Plates. — Appendices, giving Tables of Rates of Supply, Velocities, &c. &c., together with Specifications of several Works illustrated, among which will be found: Aberdeen, Bideford, Canterbury, Dundee, Halifax, Lambeth, Rotherham, Dublin, and others.

"The most systematic and valuable work upon water supply hitherto produced in English, or in any other language. . . . Mr. Humber's work is characterised almost throughout by an exhaustiveness much more distinctive of French and German than of English technical treatises."
—*Engineer.*

"We can congratulate Mr. Humber on having been able to give so large an amount of information on a subject so important as the water supply of cities and towns. The plates, fifty in number, are mostly drawings of executed works, and alone would have commanded the attention of every engineer whose practice may lie in this branch of the profession."—*Builder.*

Cast and Wrought Iron Bridge Construction.
A COMPLETE AND PRACTICAL TREATISE ON CAST AND WROUGHT IRON BRIDGE CONSTRUCTION, *including Iron Foundations*. In Three Parts—Theoretical, Practical, and Descriptive. By WILLIAM HUMBER, A-M. Inst. C.E., and M. Inst. M.E. Third Edition, Revised and much improved, with 115 Double Plates (20 of which now first appear in this edition), and numerous Additions to the Text. In Two Vols., imp. 4to, £6 16s. 6d. half-bound in morocco.

"A very valuable contribution to the standard literature of civil engineering. In addition to elevations, plans and sections, large scale details are given which very much enhance the instructive worth of these illustrations. No engineer would willingly be without so valuable a fund of information. —*Civil Engineer and Architect's Journal.*

"Mr. Humber's stately volumes, lately issued—in which the most important bridges erected during the last five years, under the direction of the late Mr. Brunel, Sir W. Cubitt, Mr. Hawkshaw, Mr. Page, Mr. Fowler, Mr. Hemans, and others among our most eminent engineers, are drawn and specified in great detail."—*Engineer.*

HUMBER'S GREAT WORK ON MODERN ENGINEERING.

Complete in Four Volumes, imperial 4to, price £12 12s., half-morocco, each Volume sold separately as follows:—

A RECORD OF THE PROGRESS OF MODERN ENGINEERING. FIRST SERIES. Comprising Civil, Mechanical, Marine, Hydraulic, Railway, Bridge, and other Engineering Works, &c. By WILLIAM HUMBER, A-M. Inst. C.E., &c. Imp. 4to, with 36 Double Plates, drawn to a large scale, Photographic Portrait of John Hawkshaw, C.E., F.R.S., &c., and copious descriptive Letterpress, Specifications, &c., £3 3s. half-morocco.

List of the Plates and Diagrams.

Victoria Station and Roof, L. B. & S. C. R. (8 plates); Southport Pier (2 plates); Victoria Station and Roof, L. C. & D. and G. W. R. (6 plates); Roof of Cremorne Music Hall; Bridge over G. N. Railway; Roof of Station, Dutch Rhenish Rail (2 plates); Bridge over the Thames, West London Extension Railway (5 plates); Armour Plates: Suspension Bridge, Thames (4 plates); The Allen Engine; Suspension Bridge, Avon (3 plates); Underground Railway (3 plates).

"Handsomely lithographed and printed. It will find favour with many who desire to preserve in a permanent form copies of the plans and specifications prepared for the guidance of the contractors for many important engineering works."—*Engineer.*

HUMBER'S RECORD OF MODERN ENGINEERING. SECOND SERIES. Imp. 4to, with 36 Double Plates, Photographic Portrait of Robert Stephenson, C.E., M.P., F.R.S., &c., and copious descriptive Letterpress, Specifications, &c., £3 3s. half-morocco.

List of the Plates and Diagrams.

Birkenhead Docks, Low Water Basin (15 plates); Charing Cross Station Roof, C. C. Railway (3 plates); Digswell Viaduct, Great Northern Railway; Robbery Wood Viaduct, Great Northern Railway; Iron Permanent Way; Clydach Viaduct, Merthyr, Tredegar, and Abergavenny Railway; Ebbw Viaduct, Merthyr, Tredegar, and Abergavenny Railway; College Wood Viaduct, Cornwall Railway; Dublin Winter Palace Roof (3 plates); Bridge over the Thames, L. C. & D. Railway (6 plates); Albert Harbour, Greenock (4 plates).

"Mr. Humber has done the profession good and true service, by the fine collection of examples he has here brought before the profession and the public."—*Practical Mechanic's Journal.*

HUMBER'S RECORD OF MODERN ENGINEERING. THIRD SERIES. Imp. 4to, with 40 Double Plates, Photographic Portrait of J. R. M'Clean, Esq., late Pres. Inst. C.E., and copious descriptive Letterpress, Specifications, &c., £3 3s. half-morocco.

List of the Plates and Diagrams.

MAIN DRAINAGE, METROPOLIS.—*North Side.*—Map showing Interception of Sewers; Middle Level Sewer (2 plates); Outfall Sewer, Bridge over River Lea (3 plates); Outfall Sewer, Bridge over Marsh Lane, North Woolwich Railway, and Bow and Barking Railway Junction; Outfall Sewer, Bridge over Bow and Barking Railway (3 plates); Outfall Sewer, Bridge over East London Waterworks' Feeder (2 plates); Outfall Sewer, Reservoir (2 plates); Outfall Sewer, Tumbling Bay and Outlet; Outfall Sewer, Penstocks. *South Side.*—Outfall Sewer, Bermondsey Branch (2 plates); Outfall Sewer, Reservoir and Outlet (4 plates); Outfall Sewer, Filth Hoist; Sections of Sewers (North and South Sides).
THAMES EMBANKMENT.—Section of River Wall; Steamboat Pier, Westminster (2 plates); Landing Stairs between Charing Cross and Waterloo Bridges; York Gate (2 plates); Overflow and Outlet at Savoy Street Sewer (3 plates); Steamboat Pier, Waterloo Bridge (3 plates); Junction of Sewers, Plans and Sections; Gullies, Plans and Sections; Rolling Stock; Granite and Iron Forts.

"The drawings have a constantly increasing value, and whoever desires to possess clear representations of the two great works carried out by our Metropolitan Board will obtain Mr. Humber's volume."—*Engineer.*

HUMBER'S RECORD OF MODERN ENGINEERING. FOURTH SERIES. Imp. 4to, with 36 Double Plates, Photographic Portrait of John Fowler, Esq., late Pres. Inst. C.E., and copious descriptive Letterpress, Specifications, &c., £3 3s. half-morocco.

List of the Plates and Diagrams.

Abbey Mills Pumping Station, Main Drainage, Metropolis (4 plates); Barrow Docks (5 plates); Manquis Viaduct, Santiago and Valparaiso Railway (2 plates); Adam's Locomotive, St. Helen's Canal Railway (2 plates); Cannon Street Station Roof, Charing Cross Railway (3 plates); Road Bridge over the River Moka (2 plates); Telegraphic Apparatus for Mesopotamia; Viaduct over the River Wye, Midland Railway (3 plates); St. Germans Viaduct, Cornwall Railway (2 plates); Wrought-Iron Cylinder for Diving Bell; Millwall Docks (6 plates); Milroy's Patent Excavator; Metropolitan District Railway (6 plates); Harbours, Ports, and Breakwaters (3 plates).

"We gladly welcome another year's issue of this valuable publication from the able pen of Mr. Humber. The accuracy and general excellence of this work are well known, while its usefulness in giving the measurements and details of some of the latest examples of engineering, as carried out by the most eminent men in the profession, cannot be too highly prized."—*Artisan.*

CIVIL ENGINEERING, SURVEYING, etc.   3

Trigonometrical Surveying.
AN OUTLINE OF THE METHOD OF CONDUCTING A TRIGONOMETRICAL SURVEY, for the Formation of Geographical and Topographical Maps and Plans, Military Reconnaissance, Levelling, &c., with Useful Problems, Formulæ, and Tables. By Lieut.-General FROME, R.E. Fourth Edition, Revised and partly Re-written by Captain CHARLES WARREN, R.E. With 19 Plates and 115 Woodcuts, royal 8vo, 16s. cloth.

"The simple fact that a fourth edition has been called for is the best testimony to its merits. No words of praise from us can strengthen the position so well and so steadily maintained by this work. Captain Warren has revised the entire work, and made such additions as were necessary to bring every portion of the contents up to the present date."—*Broad Arrow.*

Oblique Bridges.
A PRACTICAL AND THEORETICAL ESSAY ON OBLIQUE BRIDGES. With 13 large Plates. By the late GEORGE WATSON BUCK, M.I.C.E. Third Edition, revised by his Son, J. H. WATSON BUCK, M.I.C.E. and with the addition of Description to Diagrams for Facilitating the Construction of Oblique Bridges by W. H. BARLOW, M.I.C.E. Royal 8vo, 12s. cloth.

"The standard text-book for all engineers regarding skew arches is Mr. Buck's treatise, and it would be impossible to consult a better."—*Engineer.*

"Mr. Buck's treatise is recognised as a standard text-book, and his treatment has divested the subject of many of the intricacies supposed to belong to it. As a guide to the engineer and architect, on a confessedly difficult subject, Mr. Buck's work is unsurpassed."—*Building News.*

Bridge Construction in Masonry, Timber and Iron.
EXAMPLES OF BRIDGE AND VIADUCT CONSTRUCTION OF MASONRY, TIMBER, AND IRON. Consisting of 46 Plates from the Contract Drawings or Admeasurement of select Works. By W. D. HASKOLL, C.E. Second Edition, with the addition of 554 Estimates, and the Practice of Setting out Works. Illustrated with 6 pages of Diagrams. Imp. 4to, £2 12s. 6d. half-morocco.

"A work of the present nature by a man of Mr. Haskoll's experience must prove invaluable. The tables of estimates will considerably enhance its value."—*Engineering.*

Earthwork.
EARTHWORK TABLES. Showing the Contents in Cubic Yards of Embankments, Cuttings, &c., of Heights or Depths up to an average of 80 feet. By JOSEPH BROADBENT, C.E., and FRANCIS CAMPIN, C.E. Crown 8vo, 5s. cloth.

"The way in which accuracy is attained, by a simple division of each cross section into three elements, two in which are constant and one variable, is ingenious."—*Athenæum.*

Barlow's Strength of Materials, enlarged.
A TREATISE ON THE STRENGTH OF MATERIALS; with Rules for Application in Architecture, the Construction of Suspension Bridges, Railways, &c. By PETER BARLOW, F.R.S. A New Edition, revised by his Sons, P. W. BARLOW, F.R.S., and W. H. BARLOW, F.R.S.; to which are added, Experiments by HODGKINSON, FAIRBAIRN, and KIRKALDY; and Formulæ for Calculating Girders, &c. Arranged and Edited by W. HUMBER, A-M. Inst. C.E. Demy 8vo, 400 pp., with 19 large Plates and numerous Woodcuts, 18s. cloth.

'Valuable alike to the student, tyro, and the experienced practitioner, it will always rank in future as it has hitherto done, as the standard treatise on that particular subject."—*Engineer.*

There is no greater authority than Barlow."—*Building News.*

"As a scientific work of the first class, it deserves a foremost place on the bookshelves of every civil engineer and practical mechanic."—*English Mechanic.*

Strains, Formulæ and Diagrams for Calculation of.
A HANDY BOOK FOR THE CALCULATION OF STRAINS IN GIRDERS AND SIMILAR STRUCTURES, AND THEIR STRENGTH. Consisting of Formulæ and Corresponding Diagrams, with numerous details for Practical Application, &c. By WILLIAM HUMBER, A-M. Inst. C. E., &c. Fourth Edition. Crown 8vo, nearly 100 Woodcuts and 3 Plates, 7s. 6d. cloth.

The formulæ are neatly expressed, and the diagrams good."—*Athenæum.*

"We heartily commend this really handy book to our engineer and architect readers."—*English Mechanic.*

Survey Practice.

AID TO SURVEY PRACTICE, for Reference in Surveying, Levelling, Setting-out and in Route Surveys of Travellers by Land and Sea. With Tables, Illustrations, and Records. By LOWIS D'A. JACKSON, A.M.I.C.E., Author of "Hydraulic Manual," "Modern Metrology," &c. Large crown 8vo, 12s. 6d. cloth.

"Mr. Jackson has produced a valuable *vade-mecum* for the surveyor. We can recommend this book as containing an admirable supplement to the teaching of the accomplished surveyor."—*Athenæum*.

"As a text-book we should advise all surveyors to place it in their libraries, and study well the matured instructions afforded in its pages."—*Colliery Guardian*.

"The author brings to his work a fortunate union of theory and practical experience which, aided by a clear and lucid style of writing, renders the book a very useful one."—*Builder*.

Surveying, Land and Marine.

LAND AND MARINE SURVEYING, in Reference to the Preparation of Plans for Roads and Railways; Canals, Rivers, Towns' Water Supplies: Docks and Harbours. With Description and Use of Surveying Instruments. By W. DAVIS HASKOLL, C.E., Author of "Bridge and Viaduct Construction," &c. Second Edition, Revised, with Additions. Large crown 8vo, 9s. cloth.

"This book must prove of great value to the student. We have no hesitation in recommending it, feeling assured that it will more than repay a careful study."—*Mechanical World*.

"A most useful and well arranged book for the aid of a student. We can strongly recommend it as a carefully written and valuable text-book. It enjoys a well-deserved repute among surveyors."—*Builder*.

Levelling.

A TREATISE ON THE PRINCIPLES AND PRACTICE OF LEVELLING. Showing its Application to purposes of Railway and Civil Engineering, in the Construction of Roads; with Mr. TELFORD's Rules for the same. By FREDERICK W. SIMMS, F.G.S., M. Inst. C.E. Seventh Edition, with the addition of LAW's Practical Examples for Setting-out Railway Curves, and TRAUTWINE's Field Practice of Laying-out Circular Curves. With 7 Plates and numerous Woodcuts, 8vo, 8s. 6d. cloth. **** TRAUTWINE on Curves, separate, 5s.

"The text-book on levelling in most of our engineering schools and colleges."—*Engineer*.

"The publishers have rendered a substantial service to the profession, especially to the younger members, by bringing out the present edition of Mr. Simms' useful work."—*Engineering*.

Tunnelling.

PRACTICAL TUNNELLING. Explaining in detail the Setting-out of the works, Shaft-sinking and Heading-driving, Ranging the Lines and Levelling underground, Sub-Excavating, Timbering, and the Construction of the Brickwork of Tunnels, with the amount of Labour required for, and the Cost of, the various portions of the work. By FREDERICK W. SIMMS, F.G.S., M. Inst. C.E. Third Edition, Revised and Extended by D. KINNEAR CLARK, M. Inst. C.E. Imp. 8vo, with 21 Folding Plates and numerous Wood Engravings, 30s. cloth.

"The estimationi n which Mr. Simms' book on tunnelling has been held for over thirty years cannot be more truly expressed than in the words of the late Professor Rankine:—'The best source of information on the subject of tunnels is Mr. F. W. Simms' work on Practical Tunnelling.'"—*Architect*.

"It has been regarded from the first as a text-book of the subject. .... Mr. Clark has added immensely to the value of the book."—*Engineer*.

"The additional chapters by Mr. Clark, containing as they do numerous examples of modern practice, bring the book well up to date."—*Engineering*.

Statics, Graphic and Analytic.

GRAPHIC AND ANALYTIC STATICS, in Theory and Comparison: Their Practical Application to the Treatment of Stresses in Roofs, Solid Girders, Lattice, Bowstring and Suspension Bridges, Braced Iron Arches and Piers, and other Frameworks. To which is added a Chapter on Wind Pressures. By R. HUDSON GRAHAM, C.E. With numerous Examples, many taken from existing Structures. 8vo, 16s. cloth.

"Mr. Graham's book will find a place wherever graphic and analytic statics are used or studied."—*Engineer*.

"This exhaustive treatise is admirably adapted for the architect and engineer, and will tend to wean the profession from a tedious and laboured mode of calculation. To prove the accuracy of the graphical demonstrations, the author compares them with the analytic formulæ given by Rankine."—*Building News*.

"The work is excellent from a practical point of view, and has evidently been prepared with much care. It is an excellent text-book for the practical draughtsman."—*Athenæum*.

Hydraulic Tables.

HYDRAULIC TABLES, CO-EFFICIENTS, and FORMULÆ for finding the Discharge of Water from Orifices, Notches, Weirs, Pipes, and Rivers. With New Formulæ, Tables and General Information on Rainfall, Catchment-Basins, Drainage, Sewerage, Water Supply for Towns and Mill Power. By JOHN NEVILLE, Civil Engineer, M.R.I.A. Third Edition, carefully revised, with considerable Additions. Numerous Illustrations. Crown 8vo, 14s. cloth.

"Alike valuable to students and engineers in practice; its study will prevent the annoyance of avoidable failures, and assist them to select the readiest means of successfully carrying out any given work connected with hydraulic engineering."—*Mining Journal.*

"It is, of all English books on the subject, the one nearest to completion. . . . From the good arrangement of the matter, the clear explanations, and abundance of formulæ, the carefully calculated tables, and, above all, the thorough acquaintance with both theory and construction which is displayed from first to last, the book will be found to be an acquisition."—*Architect.*

River Engineering.

RIVER BARS: The Causes of their Formation, and their Treatment by "Induced Tidal Scour." With a Description of the Successful Reduction by this Method of the Bar at Dublin. By I. J. MANN, Assist. Eng. to the Dublin Port and Docks Board. Royal 8vo, 7s. 6d. cloth.

"We recommend all interested in harbour works—and, indeed, those concerned in the improvements of rivers generally—to read Mr. Mann's interesting work on the treatment of river bars."—*Engineer.*

"The author's discussion on wave-action, currents, and scour is intelligent and interesting. . . a most valuable contribution to the history of this branch of engineering."—*Engineering and Mining Journal.*

Hydraulics.

HYDRAULIC MANUAL. Consisting of Working Tables and Explanatory Text. Intended as a Guide in Hydraulic Calculations and Field Operations. By LOWIS D'A. JACKSON. Fourth Edition. Rewritten and Enlarged. Large crown 8vo, 16s. cloth.

"From the great mass of material at his command the author has constructed a manual which may be accepted as a trustworthy guide to this branch of the engineer's profession. We can heartily recommend this volume to all who desire to be acquainted with the latest development of this important subject."—*Engineering.*

"The standard work in this department of mechanics. The present edition has been brought abreast of the most recent practice."—*Scotsman.*

"The most useful feature of this work is its freedom from what is superannuated and its thorough adoption of recent experiments; the text is, in fact, in great part a short account of the great modern experiments."—*Nature.*

Tramways and their Working.

TRAMWAYS: THEIR CONSTRUCTION AND WORKING. Embracing a Comprehensive History of the System; with an exhaustive Analysis of the various Modes of Traction, including Horse-Power, Steam, Heated Water, and Compressed Air; a Description of the Varieties of Rolling Stock; and ample Details of Cost and Working Expenses: the Progress recently made in Tramway Construction, &c. &c. By D. KINNEAR CLARK, M. Inst. C.E. With over 200 Wood Engravings, and 13 Folding Plates. Two Vols., large crown 8vo, 30s. cloth.

"All interested in tramways must refer to it, as all railway engineers have turned to the author's work 'Railway Machinery.'"—*Engineer.*

"An exhaustive and practical work on tramways, in which the history of this kind of locomotion, and a description and cost of the various modes of laying tramways, are to be found."—*Building News.*

"The best form of rails, the best mode of construction, and the best mechanical appliances are so fairly indicated in the work under review, that any engineer about to construct a tramway will be enabled at once to obtain the practical information which will be of most service to him."—*Athenæum.*

Oblique Arches.

A PRACTICAL TREATISE ON THE CONSTRUCTION OF OBLIQUE ARCHES. By JOHN HART. Third Edition, with Plates. Imperial 8vo, 8s. cloth.

Strength of Girders.

GRAPHIC TABLE FOR FACILITATING THE COMPUTATION OF THE WEIGHTS OF WROUGHT IRON AND STEEL GIRDERS, &c., for Parliamentary and other Estimates. By J. H. WATSON BUCK, M. Inst. C.E. On a Sheet, 2s.6d.

Tables for Setting-out Curves.

TABLES OF TANGENTIAL ANGLES AND MULTIPLES for Setting-out Curves from 5 to 200 Radius. By ALEXANDER BEAZELEY, M. Inst. C.E. Third Edition. Printed on 48 Cards, and sold in a cloth box, waistcoat-pocket size, 3s. 6d.

"Each table is printed on a small card, which, being placed on the theodolite, leaves the hands free to manipulate the instrument—no small advantage as regards the rapidity of work."—*Engineer*.

"Very handy; a man may know that all his day's work must fall on two of these cards, which he puts into his own card-case, and leaves the rest behind."—*Athenæum*.

Engineering Fieldwork.

THE PRACTICE OF ENGINEERING FIELDWORK, applied to Land and Hydraulic, Hydrographic, and Submarine Surveying and Levelling. Second Edition, Revised, with considerable Additions, and a Supplement on Waterworks, Sewers, Sewage, and Irrigation. By W. DAVIS HASKOLL, C.E. Numerous Folding Plates. In One Volume, demy 8vo, £1 5s. cloth.

Large Tunnel Shafts.

THE CONSTRUCTION OF LARGE TUNNEL SHAFTS: A Practical and Theoretical Essay. By J. H. WATSON BUCK, M. Inst. C.E., Resident Engineer, London and North-Western Railway. Illustrated with Folding Plates, royal 8vo, 12s. cloth.

"Many of the methods given are of extreme practical value to the mason; and the observations on the form of arch, the rules for ordering the stone, and the construction of the templates will be found of considerable use. We commend the book to the engineering profession."—*Building News*.

"Will be regarded by civil engineers as of the utmost value, and calculated to save much time and obviate many mistakes."—*Colliery Guardian*.

Field-Book for Engineers.

THE ENGINEER'S, MINING SURVEYOR'S, AND CONTRACTOR'S FIELD-BOOK. Consisting of a Series of Tables, with Rules, Explanations of Systems, and use of Theodolite for Traverse Surveying and Plotting the Work with minute accuracy by means of Straight Edge and Set Square only; Levelling with the Theodolite, Casting-out and Reducing Levels to Datum, and Plotting Sections in the ordinary manner; setting-out Curves with the Theodolite by Tangential Angles and Multiples, with Right and Left-hand Readings of the Instrument: Setting-out Curves without Theodolite, on the System of Tangential Angles by sets of Tangents and Offsets: and Earthwork Tables to 80 feet deep, calculated for every 6 inches in depth. By W. DAVIS HASKOLL, C.E. With numerous Woodcuts. Fourth Edition, Enlarged. Crown 8vo, 12s. cloth.

"The book is very handy, and the author might have added that the separate tables of sines and tangents to every minute will make it useful for many other purposes, the genuine traverse tables existing all the same."—*Athenæum*.

"Every person engaged in engineering field operations will estimate the importance of such a work and the amount of valuable time which will be saved by reference to a set of reliable tables prepared with the accuracy and fulness of those given in this volume."—*Railway News*.

Earthwork, Measurement and Calculation of.

A MANUAL ON EARTHWORK. By ALEX. J. S. GRAHAM, C.E. With numerous Diagrams. 18mo, 2s. 6d. cloth.

"A great amount of practical information, very admirably arranged, and available for rough estimates, as well as for the more exact calculations required in the engineer's and contractor's offices."—*Artizan*.

Strains.

THE STRAINS ON STRUCTURES OF IRONWORK; with Practical Remarks on Iron Construction. By F. W. SHEILDS, M. Inst. C.E. Second Edition, with 5 Plates. Royal 8vo, 5s. cloth.

"The student cannot find a better little book on this subject."—*Engineer*.

Strength of Cast Iron, etc.

A PRACTICAL ESSAY ON THE STRENGTH OF CAST IRON AND OTHER METALS. By THOMAS TREDGOLD, C.E. Fifth Edition, including HODGKINSON's Experimental Researches. 8vo, 12s. cloth.

MECHANICS & MECHANICAL ENGINEERING.

The Modernised "Templeton."

THE PRACTICAL MECHANIC'S WORKSHOP COMPANION. Comprising a great variety of the most useful Rules and Formulæ in Mechanical Science, with numerous Tables of Practical Data and Calculated Results for Facilitating Mechanical Operations. By WILLIAM TEMPLETON, Author of "The Engineer's Practical Assistant," &c. &c. An Entirely New Edition, Revised, Modernised, and considerably Enlarged by WALTER S. HUTTON, C.E., Author of "The Works' Manager's Handbook of Modern Rules, Tables, and Data for Engineers," &c. Fcap. 8vo, nearly 500 pp., with 8 Plates and upwards of 250 Illustrative Diagrams, 6s., strongly bound for workshop or pocket wear and tear. [*Just published.*

☞ TEMPLETON'S "MECHANIC'S WORKSHOP COMPANION" has been for more than a quarter of a century deservedly popular, having run through numerous Editions; and, as a recognised Text-Book and well-worn and thumb-marked vade mecum of several generations of intelligent and aspiring workmen, it has had the reputation of having been the means of raising many of them in their position in life.

In its present greatly Enlarged, Improved and Modernised form, the Publishers are sure that it will commend itself to the English workmen of the present day all the world over, and become, like its predecessors, their indispensable friend and referee.

A smaller type having been adopted, and the page increased in size, while the number of pages has advanced from about 330 to nearly 500, the book practically contains double the amount of matter that was comprised in the original work.

** OPINIONS OF THE PRESS.

"In its modernised form Hutton's 'Templeton' should have a wide sale, fo it contains much valuable information which the mechanic will often find of use, and not a few tables and notes which he might look for in vain in other works. This modernised edition will be appreciated by all who have learned value the original editions of 'Templeton.'"—*English Mechanic.*

"It has met with great success in the engineering workshop, as we can testify; and there are a great many men who, in a great measure, owe their rise in life to this little book."—*Building News.*

Engineer's and Machinist's Assistant.

THE ENGINEER'S, MILLWRIGHT'S, and MACHINIST'S PRACTICAL ASSISTANT. A collection of Useful Tables, Rules and Data. By WILLIAM TEMPLETON. Seventh Edition, with Additions. 18mo, 2s. 6d. cloth.

"Templeton's handbook occupies a foremost place among books of this kind. A more suitable present to an apprentice to any of the mechanical trades could not possibly be made."—*Building News.*

Turning.

LATHE-WORK: A Practical Treatise on the Tools, Appliances, and Processes employed in the Art of Turning. By PAUL N. HASLUCK. Third Edition, Revised and Enlarged. Crown 8vo, 5s. cloth.

"Written by a man who knows, not only how work ought to be done, but who also knows how to do it, and how to convey his knowledge to others."—*Engineering.*

"We can safely recommend the work to young engineers. To the amateur it will simply be invaluable. To the student it will convey a great deal of useful information."—*Engineer.*

"A compact, succinct, and handy guide to lathe-work did not exist in our language until Mr. Hasluck, by the publication of this treatise, gave the turner a true *vade-mecum.*"—*House Decorator.*

Iron and Steel.

"IRON AND STEEL": A Work for the Forge, Foundry, Factory, and Office. Containing ready, useful, and trustworthy Information for Ironmasters and their Stock-takers; Managers of Bar, Rail, Plate, and Sheet Rolling Mills; Iron and Metal Founders; Iron Ship and Bridge Builders; Mechanical, Mining, and Consulting Engineers; Architects, Contractors, Builders, and Professional Draughtsmen. By CHARLES HOARE, Author of "The Slide Rule," &c. Eighth Edition, Revised throughout and considerably Enlarged. With folding Scales of "Foreign Measures compared with the English Foot," and "Fixed Scales of Squares, Cubes, and Roots, Areas, Decimal Equivalents, &c." Oblong 32mo, leather, elastic band, 6s.

"For comprehensiveness the book has not its equal."—*Iron.*

"One of the best of the pocket books, and a useful companion in other branches of work than iron and steel."—*English Mechanic.*

"We cordially recommend this book to those engaged in considering the details of all kinds of iron and steel works."—*Naval Science.*

Stone-working Machinery.

STONE-WORKING MACHINERY, and the Rapid and Economical Conversion of Stone. With Hints on the Arrangement and Management of Stone Works. By M. POWIS BALE, M.I.M.E., A.M.I.C.E. With numerous Illustrations. Large crown 8vo, 9s. cloth.

"The book should be in the hands of every mason or student of stone-work."—*Colliery Guardian.*

Engineer's Reference Book.

THE WORKS' MANAGER'S HANDBOOK OF MODERN RULES, TABLES, AND DATA. For Engineers, Millwrights, and Boiler Makers; Tool Makers, Machinists, and Metal Workers; Iron and Brass Founders, &c. By W. S. HUTTON, Civil and Mechanical Engineer. Third Edition, carefully revised, with Additions. In One handsome Volume, medium 8vo, price 15s. strongly bound.

"The author treats every subject from the point of view of one who has collected workshop notes for application in workshop practice, rather than from the theoretical or literary aspect. The volume contains a great deal of that kind of information which is gained only by practical experience, and is seldom written in books."—*Engineer.*

"The volume is an exceedingly useful one, brimful with engineers notes, memoranda, and rules, and well worthy of being on every mechanical engineer's bookshelf. . . . There is valuable information on every page."—*Mechanical World.*

"The information is precisely that likely to be required in practice. . . . The work forms a desirable addition to the library, not only of the works' manager, but of anyone connected with general engineering."—*Mining Journal.*

"A formidable mass of facts and figures, readily accessible through an elaborate index . . . . Such a volume will be found absolutely necessary as a book of reference in all sorts of 'works' connected with the metal trades. . . . Any ordinary foreman or workman can find all he wants in the crowded pages of this useful work."—*Ryland's Iron Trades Circular*

Engineering Construction.

PATTERN-MAKING : A Practical Treatise, embracing the Main Types of Engineering Construction, and including Gearing, both Hand and Machine made, Engine Work, Sheaves and Pulleys, Pipes and Columns, Screws, Machine Parts, Pumps and Cocks, the Moulding of Patterns in Loam and Greensand, &c., together with the methods of Estimating the weight of Castings; to which is added an Appendix of Tables for Workshop Reference. By a FOREMAN PATTERN MAKER. With upwards of Three Hundred and Seventy Illustrations. Crown 8vo, 7s. 6d. cloth.

"A well-written technical guide, evidently written by a man who understands and has practised what he has written about; he says what he has to say in a plain, straightforward manner. We cordially recommend the treatise to engineering students, young journeymen, and others desirous of being initiated into the mysteries of pattern-making."—*Builder.*

"We can confidently recommend this comprehensive treatise."—*Building News.*

"A valuable contribution to the literature of an important branch of engineering construction, which is likely to prove a welcome guide to many workmen, especially to draughtsmen who have lacked a training in the shops, pupils pursuing their practical studies in our factories, and to employers and managers in engineering works."—*Hardware Trade Journal.*

"More than 370 illustrations help to explain the text, which is, however, always clear and explicit, thus rendering the work an excellent *vade mecum* for the apprentice who desires to become master of his trade."—*English Mechanic.*

Smith's Tables for Mechanics, etc.

TABLES, MEMORANDA, AND CALCULATED RESULTS, FOR MECHANICS, ENGINEERS, ARCHITECTS, BUILDERS, etc. Selected and Arranged by FRANCIS SMITH. Third Edition, Revised and Enlarged, 250 pp., waistcoat-pocket size, 1s. 6d. limp leather.

"It would, perhaps, be as difficult to make a small pocket-book selection of notes and formulæ to suit ALL engineers as it would be to make a universal medicine; but Mr. Smith's waistcoat-pocket collection may be looked upon as a successful attempt."—*Engineer.*

"The best example we have ever seen of 250 pages of useful matter packed into the dimensions of a card-case."—*Building News.*

"A veritable pocket treasury of knowledge."—*Iron.*

The High-Pressure Steam Engine.

THE HIGH-PRESSURE STEAM-ENGINE : An Exposition of its Comparative Merits and an Essay towards an Improved System of Construction. By Dr. ERNST ALBAN. Translated from the German, with Notes, by Dr. POLE, M. Inst. C.E., &c. With 28 Plates. 8vo, 16s. 6d. cloth.

"Goes thoroughly into the examination of the high-pressure engine, the boiler, and its appendages, and deserves a place in every scientific library."—*Steam Shipping Chronicle.*

Steam Boilers.

A TREATISE ON STEAM BOILERS: Their Strength, Construction, and Economical Working. By ROBERT WILSON, C.E. Fifth Edition. 12mo, 6s. cloth.

"The best treatise that has ever been published on steam boilers."—*Engineer.*

"The author shows himself perfect master of his subject, and we heartily recommend all employing steam power to possess themselves of the work."—*Ryland's Iron Trade Circular.*

Boiler Making.

THE BOILER-MAKER'S READY RECKONER. With Examples of Practical Geometry and Templating, for the Use of Platers, Smiths and Riveters. By JOHN COURTNEY, Edited by D. K. CLARK, M.I.C.E. Second Edition, revised, with Additions, 12mo, 5s. half-bound.

"A most useful work. .... No workman or apprentice should be without this book. — *Iron Trade Circular.*

"A reliable guide to the working boiler-maker."—*Iron.*

"Boiler-makers will readily recognise the value of this volume. . . . The tables are clearly printed, and so arranged that they can be referred to with the greatest facility, so that it cannot be doubted that they will be generally appreciated and much used."—*Mining Journal.*

Steam Engine.

TEXT-BOOK ON THE STEAM ENGINE. By T. M. GOODEVE, M.A., Barrister-at-Law, Author of "The Elements of Mechanism," &c. Seventh Edition. With numerous Illustrations. Crown 8vo, 6s. cloth.

"Professor Goodeve has given us a treatise on the steam engine which will bear comparison with anything written by Huxley or Maxwell, and we can award it no higher praise."—*Engineer.*

Steam.

THE SAFE USE OF STEAM. Containing Rules for Unprofessional Steam-users. By an ENGINEER. Fifth Edition. Sewed, 6d.

"If steam-users would but learn this little book by heart, boiler explosions would become sensations by their rarity."—*English Mechanic.*

Coal and Speed Tables.

A POCKET BOOK OF COAL AND SPEED TABLES, for Engineers and Steam-users. By NELSON FOLEY, Author of "Boiler Construction." Pocket-size, 3s. 6d. cloth; 4s. leather.

"This is a very useful book, containing very useful tables. The results given are well chosen, and the volume contains evidence that the author really understands his subject. We can recommend the work with pleasure."—*Mechanical World.*

"These tables are designed to meet the requirements of every-day use; they are of sufficient scope for most practical purposes, and may be commended to engineers and users of steam."—*Iron.*

"This pocket-book well merits the attention of the practical engineer. Mr. Foley has compiled a very useful set of tables, the information contained in which is frequently required by engineers, coal consumers and users of steam."—*Iron and Coal Trades Review.*

Fire Engineering.

FIRES, FIRE-ENGINES, AND FIRE-BRIGADES. With a History of Fire-Engines, their Construction, Use, and Management; Remarks on Fire-Proof Buildings, and the Preservation of Life from Fire; Statistics of the Fire Appliances in English Towns; Foreign Fire Systems; Hints on Fire Brigades, &c. &c. By CHARLES F. T. YOUNG, C.E. With numerous Illustrations, 544 pp., demy 8vo, £1 4s. cloth.

"To such of our readers as are interested in the subject of fires and fire apparatus, we can most heartily commend this book. It is really the only English work we now have upon the subject."—*Engineering.*

"It displays much evidence of careful research; and Mr. Young has put his facts neatly together. It is evident enough that his acquaintance with the practical details of the construction of steam fire engines, old and new, and the conditions with which it is necessary hey should comply, s accurate and full."—*Engineer.*

Gas Lighting.

COMMON SENSE FOR GAS-USERS: A Catechism of Gas-Lighting for Householders, Gasfitters, Millowners, Architects, Engineers, etc. By ROBERT WILSON, C.E., Author of "A Treatise on Steam Boilers." Second Edition. Crown 8vo, sewed, with Folding Plates and Wood Engravings, 2s. 6d.

"All gas-users will decidedly benefit, both in pocket and comfort, if they will avail themselves of Mr. Wilson's counsels."—*Engineering.*

THE POPULAR WORKS OF MICHAEL REYNOLDS
(Known as "THE ENGINE DRIVER'S FRIEND").

Locomotive-Engine Driving.
LOCOMOTIVE-ENGINE DRIVING : A Practical Manual for Engineers in charge of Locomotive Engines. By MICHAEL REYNOLDS, Member of the Society of Engineers, formerly Locomotive Inspector L. B. and S. C. R. Seventh Edition. Including a KEY TO THE LOCOMOTIVE ENGINE. With Illustrations and Portrait of Author. Crown 8vo, 4s. 6d. cloth.

"Mr. Reynolds has supplied a want, and has supplied it well. We can confidently recommend the book, not only to the practical driver, but to everyone who takes an interest in the performance of locomotive engines."—*The Engineer.*

"Were the cautions and rules given in the book to become part of the every-day working of our engine-drivers, we might have fewer distressing accidents to deplore."—*Scotsman.*

The Engineer, Fireman, and Engine-Boy.
THE MODEL LOCOMOTIVE ENGINEER, FIREMAN, and ENGINE-BOY. Comprising a Historical Notice of the Pioneer Locomotive Engines and their Inventors, with a project for the establishment of Certificates of Qualification in the Running Service of Railways. By MICHAEL REYNOLDS, Author of "Locomotive-Engine Driving." With numerous Illustrations and a fine Portrait of George Stephenson. Crown 8vo, 4s. 6d. cloth.

"From the technical knowledge of the author it will appeal to the railway man of to-day more forcibly than anything written by Dr. Smiles. . . . The volume contains information of a technical kind, and facts that every driver should be familiar with."—*English Mechanic.*

"We should be glad to see this book in the possession of everyone in the kingdom who has ever laid, or is to lay, hands on a locomotive engine."—*Iron.*

Stationary Engine Driving.
STATIONARY ENGINE DRIVING : A Practical Manual for Engineers in charge of Stationary Engines. By MICHAEL REYNOLDS. Third Edition, Enlarged. With Plates and Woodcuts. Crown 8vo, 4s. 6d. cloth.

"The author is thoroughly acquainted with his subjects, and his advice on the various points treated is clear and practical. . . . He has produced a manual which is an exceedingly useful one for the class for whom it is specially intended."—*Engineering.*

"Our author leaves no stone unturned. He is determined that his readers shall not only know something about the stationary engine, but all about it."—*Engineer.*

Continuous Railway Brakes.
CONTINUOUS RAILWAY BRAKES : A Practical Treatise on the several Systems in Use in the United Kingdom; their Construction and Performance. With copious Illustrations and numerous Tables. By MICHAEL REYNOLDS. Large crown 8vo, 9s. cloth.

"A popular explanation of the different brakes. It will be of great assistance in forming public opinion, and will be studied with benefit by those who take an interest in the brake."—*English Mechanic.*

"Written with sufficient technical detail to enable the principle and relative connection of the various parts of each particular brake to be readily grasped."—*Mechanical World.*

Engine-Driving Life.
ENGINE-DRIVING LIFE; or, Stirring Adventures and Incidents in the Lives of Locomotive-Engine Drivers. By MICHAEL REYNOLDS. Ninth Thousand. Crown 8vo, 2s. cloth.

"The book from first to last is perfectly fascinating. Wilkie Collins' most thrilling conceptions are thrown into the shade by true incidents, endless in their variety, related in every page."—*North British Mail.*

"Anyone who wishes to get a real insight into railway life cannot do better than read 'Engine-Driving Life' for himself; and if he once take it up he will find that the author's enthusiasm and real love of the engine-driving profession will carry him on till he has read every page."—*Saturday Review.*

Pocket Companion for Enginemen.
THE ENGINEMAN'S POCKET COMPANION AND PRACTICAL EDUCATOR FOR ENGINEMEN, BOILER ATTENDANTS, AND MECHANICS. By MICHAEL REYNOLDS, Mem. S. E., Author of "Locomotive Engine-Driving," "Stationary Engine-Driving," &c. With Forty-five Illustrations and numerous Diagrams. Royal 18mo, 3s. 6d., strongly bound in cloth for pocket wear. [*Just published.*

ARCHITECTURE, BUILDING, etc.

Construction.
THE SCIENCE OF BUILDING: An Elementary Treatise on the Principles of Construction. By E. WYNDHAM TARN, M.A., Architect. Second Edition, Revised, with 58 Engravings. Crown 8vo, 7s. 6d. cloth.

"A very valuable book, which we strongly recommend to all students."—*Builder.*

"No architectural student should be without this handbook of constructional knowledge."—*Architect.*

Villa Architecture.
A HANDY BOOK OF VILLA ARCHITECTURE: Being a Series of Designs for Villa Residences in various Styles. With Outline Specifications and Estimates. By C. WICKES, Architect, Author of "The Spires and Towers of England," &c. 30 Plates, 4to, half-morocco, gilt edges, £1 1s.

*** Also an Enlarged Edition of the above. 61 Plates, with Outline Specifications, Estimates, &c. £2 2s. half-morocco.

"The whole of the designs bear evidence of their being the work of an artistic architect, and they will prove very valuable and suggestive."—*Building News.*

Useful Text-Book for Architects.
THE ARCHITECT'S GUIDE: Being a Text-Book of Useful Information for Architects, Engineers, Surveyors, Contractors, Clerks of Works, &c. &c. By FREDERICK ROGERS, Architect, Author of "Specifications for Practical Architecture," &c. Second Edition, Revised and Enlarged. With numerous Illustrations. Crown 8vo, 6s. cloth.

"As a text-book of useful information for architects, engineers, surveyors, &c., it would be hard to find a handier or more complete little volume."—*Standard.*

"A young architect could hardly have a better guide-book."—*Timber Trades Journal.*

Taylor and Cresy's Rome.
THE ARCHITECTURAL ANTIQUITIES OF ROME. By the late G. L. TAYLOR, Esq., F.R.I.B.A., and EDWARD CRESY, Esq. New Edition, thoroughly revised by the Rev. ALEXANDER TAYLOR, M.A. (son of the late G. L. Taylor, Esq.), Fellow of Queen's College, Oxford, and Chaplain of Gray's Inn. Large folio, with 130 Plates, half-bound, £3 3s.

N.B.—This is the only book which gives on a large scale, and with the precision of architectural measurement, the principal Monuments of Ancient Rome in plan, elevation, and detail.

"Taylor and Cresy's work has from its first publication been ranked among those professional books which cannot be bettered. . . . It would be difficult to find examples of drawings, even among those of the most painstaking students of Gothic, more thoroughly worked out than are the one hundred and thirty plates in this volume."—*Architect.*

Drawing for Builders and Students in Architecture.
PRACTICAL RULES ON DRAWING, for the Operative Builder and Young Student in Architecture. By GEORGE PYNE. With 14 Plates, 4to, 7s. 6d. boards.

Civil Architecture.
THE DECORATIVE PART OF CIVIL ARCHITECTURE. By Sir WILLIAM CHAMBERS, F.R.S. With Illustrations, Notes, and an Examination of Grecian Architecture, by JOSEPH GWILT, F.S.A. Edited by W. H. LEEDS. 66 Plates, 4to, 21s. cloth.

The House-Owner's Estimator.
THE HOUSE-OWNER'S ESTIMATOR; or, What will it Cost to Build, Alter, or Repair? A Price Book adapted to the Use of Unprofessional People, as well as for the Architectural Surveyor and Builder. By the late JAMES D. SIMON, A.R.I.B.A. Edited and Revised by FRANCIS T. W. MILLER, A.R.I.B.A. With numerous Illustrations. Third Edition, Revised. Crown 8vo, 3s. 6d. cloth.

"In two years it will repay its cost a hundred times over."—*Field.*

"A very handy book,"—*English Mechanic.*

Designing, Measuring, and Valuing.

THE STUDENT'S GUIDE to the PRACTICE of MEASURING AND VALUING ARTIFICERS' WORKS. Containing Directions for taking Dimensions, Abstracting the same, and bringing the Quantities into Bill, with Tables of Constants, and copious Memoranda for the Valuation of Labour and Materials in the respective Trades of Bricklayer and Slater, Carpenter and Joiner, Painter and Glazier, Paperhanger, &c. With 8 Plates and 63 Woodcuts. Originally edited by EDWARD DOBSON, Architect. Fifth Edition, Revised, with considerable Additions on Mensuration and Construction, and a New Chapter on Dilapidations, Repairs, and Contracts, by E. WYNDHAM TARN, M.A. Crown 8vo, 9s. cloth.

"Well fulfils the promise of its title-page, and we can thoroughly recommend it to the class for whose use it has been compiled. Mr. Tarn's additions and revisions have much increased the usefulness of the work, and have especially augmented its value to students."—*Engineering*.

"The work has been carefully revised and edited by Mr. E. Wyndham Tarn, M.A., and comprises several valuable additions on construction, mensuration, dilapidations and repairs, and other matters. . . . This edition will be found the most complete treatise on the principles of measuring and valuing artificers' work that has yet been published."—*Building News*.

Pocket Estimator.

THE POCKET ESTIMATOR for the BUILDING TRADES. Being an Easy Method of Estimating the various parts of a Building collectively, more especially applied to Carpenters' and Joiners' work. By A. C. BEATON, Author of "Quantities and Measurements." Third Edition, carefully revised, 33 Woodcuts, leather, waistcoat-pocket size, 1s. 6d.

"Contains a good deal of information not easily to be obtained from the ordinary price books. The prices given are accurate, and up to date."—*Building News*.

Builder's and Surveyor's Pocket Technical Guide.

THE POCKET TECHNICAL GUIDE AND MEASURER FOR BUILDERS AND SURVEYORS. Containing a Complete Explanation of the Terms used in Building Construction, Memoranda for Reference, Technical Directions for Measuring Work in all the Building Trades, with a Treatise on the Measurement of Timber, Complete Specifications, &c. &c. By A. C. BEATON. Second Edition, with 19 Woodcuts, leather, waistcoat-pocket size, 1s. 6d.

"An exceedingly handy pocket companion, thoroughly reliable."—*Builder's Weekly Reporter*.

"This neat little compendium contains all that is requisite in carrying out contracts for excavating, tiling, bricklaying, paving, &c."—*British Trade Journal*.

Donaldson on Specifications.

THE HANDBOOK OF SPECIFICATIONS; or, Practical Guide to the Architect, Engineer, Surveyor, and Builder, in drawing up Specifications and Contracts for Works and Constructions. Illustrated by Precedents of Buildings actually executed by eminent Architects and Engineers. By Professor T. L. DONALDSON, P.R.I.B.A., &c. New Edition, in One large Vol., 8vo, with upwards of 1,000 pages of Text, and 33 Plates, £1 11s. 6d. cloth.

"In this work forty-four specifications of executed works are given, including the specifications for parts of the new Houses of Parliament, by Sir Charles Barry, and for the new Royal Exchange, by Mr. Tite, M.P. The latter, in particular, is a very complete and remarkable document. It embodies, to a great extent, as Mr. Donaldson mentions, 'the bill of quantities with the description of the works.' . . . It is valuable as a record, and more valuable still as a book of precedents. . . . Suffice it to say that Donaldson's 'Handbook of Specifications' must be bought by all architects."—*Builder*.

Bartholomew and Rogers' Specifications.

SPECIFICATIONS FOR PRACTICAL ARCHITECTURE: A Guide to the Architect, Engineer, Surveyor, and Builder; with an Essay on the Structure and Science of Modern Buildings. Upon the Basis of the Work by ALFRED BARTHOLOMEW, thoroughly Revised, Corrected, and greatly added to by FREDERICK ROGERS, Architect. Second Edition, Revised, with Additions. With numerous Illusts., medium 8vo, 15s. cloth. [*Just published*.

"The collection of specifications prepared by Mr. Rogers on the basis of Bartholomew's work is too well known to need any recommendation from us. It is one of the books with which every young architect must be equipped ; for time has shown that the specifications cannot be set aside through any defect in them."—*Architect*.

"Good forms for specifications are of considerable value, and it was an excellent idea to compile a work on the subject upon the basis of the late Alfred Bartholomew's valuable work. The second edition of Mr. Rogers's book is evidence of the want of a book dealing with modern requirements and materials."—*Building News*.

DECORATIVE ARTS, etc.

Woods and Marbles (Imitation of).

SCHOOL OF PAINTING FOR THE IMITATION OF WOODS AND MARBLES, as Taught and Practised by A. R. VAN DER BURG and P. VAN DER BURG, Directors of the Rotterdam Painting Institution. Second and Cheaper Edition. Royal folio, 18½ by 12½ in., Illustrated with 24 full-size Coloured Plates; also 12 plain Plates, comprising 154 Figures, price £1 11s. 6d.

List of Contents

Introductory Chapter — Tools required for Wood Painting—Observations on the different species of Wood: Walnut—Observations on Marble in general—Tools required for Marble Painting—St. Remi Marble: Preparation of the Paints; Process of Working—Wood Graining: Preparation of Stiff and Flat Brushes: Sketching different Grains and Knots: Glazing of Wood—Ash: Painting of Ash—Breche (Brecia) Marble: Breche Violette: Process of Working—Maple: Process of Working—The different species of White Marble: Methods of Working: Painting White Marble with Lac-dye: Painting White Marble with Poppy-paint—Mahogany:

Methods of Working—Yellow Sienna Marble: Process of Working—Juniper: Characteristics of the Natural Wood: Method of Imitation—Vert de Mer Marble: Description of the Marble: Process of Working—Oak: Description of the varieties of Oak: Manipulation of Oak-painting: Tools employed: Method of Working—Waulsort Marble: Varieties of the Marble: Process of Working—The Painting of Iron with Red Lead: How to make Putty: Out-door Work: Varnishing: Priming and Varnishing Woods and Marbles: Painting in General: Ceilings and Walls: Gilding: Transparencies, Flags, &c.

List of Plates.

1. Various Tools required for Wood Painting—2, 3. Walnut: Preliminary Stages of Graining and Finished Specimen — 4. Tools used for Marble Painting and Method of Manipulation—5, 6. St. Remi Marble: Earlier Operations and Finished Specimen—7. Methods of Sketching different Grains, Knots, &c.—8, 9. Ash: Preliminary Stages and Finished Specimen—10. Methods of Sketching Marble Grains—11, 12. Breche Marble: Preliminary Stages of Working and Finished Specimen—13. Maple: Methods of Producing the different Grains—14, 15. Bird's-eye Maple: Preliminary Stages and Finished Specimen—16. Methods of Sketching the different Species of White Marble—17, 18. White Marble: Preliminary Stages of Process and

Finished Specimen—19. Mahogany: Specimens of various Grains and Methods of Manipulation —20, 21. Mahogany: Earlier Stages and Finished Specimen—22, 23, 24. Sienna Marble: Varieties of Grain, Preliminary Stages and Finished Specimen—25, 26, 27. Juniper Wood: Methods of producing Grain, &c.: Preliminary Stages and Finished Specimen—28, 29, 30. Vert de Mer Marble: Varieties of Grain and Methods of Working Unfinished and Finished Specimens—31, 32, 33. Oak: Varieties of Grain, Tools Employed, and Methods of Manipulation, Preliminary Stages and Finished Specimen—34, 35, 36. Waulsort Marble: Varieties of Grain, Unfinished and Finished Specimens.

"Those who desire to attain skill in the art of painting woods and marbles, will find advantage in consulting this book. . . . Some of the Working Men's Clubs should give their young men the opportunity to study it."—*Builder*.

"A comprehensive guide to the art. The explanations of the processes, the manipulation and management of the colours, and the beautifully executed plates will not be the least valuable to the student who aims at making his work a faithful transcript of nature."—*Building News*.

Colour.

A GRAMMAR OF COLOURING. Applied to Decorative Painting and the Arts. By GEORGE FIELD. New Edition, adapted to the use of the Ornamental Painter and Designer. By ELLIS A. DAVIDSON. With New Coloured Diagrams and Engravings. 12mo, 3s. 6d. cloth boards.

"The book is a most useful *resumé* of the properties of pigments."—*Builder*.

House Decoration.

ELEMENTARY DECORATION. A Guide to the Simpler Forms of Everyday Art, as applied to the Interior and Exterior Decoration of Dwelling Houses, &c. By JAMES W. FACEY. With 68 Cuts. 2s. cloth limp.

"As a technical guide-book to the decorative painter it will be found reliable."—*Building News*.

*** By the same Author, just published.

PRACTICAL HOUSE DECORATION: A Guide to the Art of Ornamental Painting, the Arrangement of Colours in Apartments, and the principles of Decorative Design. With some Remarks upon the Nature and Properties of Pigments. With numerous Illustrations. 12mo, 2s. 6d. cl. limp

N.B.—*The above Two Works together in One Vol., strongly half-bound, 5s.*

House Painting, etc.

HOUSE PAINTING, GRAINING, MARBLING, AND SIGN WRITING, A Practical Manual of. By ELLIS A. DAVIDSON. Fourth Edition. With Coloured Plates and Wood Engravings. 12mo, 6s. cloth boards.

"A mass of information, of use to the amateur and of value to the practical man."—*English Mechanic*.

DELAMOTTES' WORKS on ILLUMINATION & ALPHABETS.

A PRIMER OF THE ART OF ILLUMINATION, *for the Use of Beginners:* with a Rudimentary Treatise on the Art, Practical Directions for its exercise, and Examples taken from Illuminated MSS., printed in Gold and Colours. By F. DELAMOTTE. New and cheaper edition. Small 4to, 6s. ornamental boards.

". . . . The examples of ancient MSS. recommended to the student, which, with much good sense, the author chooses from collections accessible to all, are selected with judgment and knowledge, as well as taste."—*Athenæum.*

ORNAMENTAL ALPHABETS, *Ancient and Mediæval, from the Eighth Century, with Numerals;* including Gothic, Church-Text, large and small, German, Italian, Arabesque, Initials for Illumination, Monograms Crosses, &c. &c., for the use of Architectural and Engineering Draughtsmen, Missal Painters, Masons, Decorative Painters, Lithographers, Engravers, Carvers, &c. &c. Collected and Engraved by F. DELAMOTTE, and printed in Colours. New and Cheaper Edition. Royal 8vo, oblong, 2s. 6d. ornamental boards.

" For those who insert enamelled sentences round gilded chalices, who blazon shop legends over shop-doors, who letter church walls with pithy sentences from the Decalogue, this book will be useful."—*Athenæum.*

EXAMPLES OF MODERN ALPHABETS, *Plain and Ornamental;* including German, Old English, Saxon, Italic, Perspective, Greek, Hebrew, Court Hand, Engrossing, Tuscan, Riband, Gothic, Rustic, and Arabesque; with several Original Designs, and an Analysis of the Roman and Old English Alphabets, large and small, and Numerals, for the use of Draughtsmen, Surveyors, Masons, Decorative Painters, Lithographers, Engravers, Carvers, &c. Collected and Engraved by F. DELAMOTTE, and printed in Colours. New and Cheaper Edition. Royal 8vo, oblong, 2s. 6d. ornamental boards.

"There is comprised in it every possible shape into which the letters of the alphabet and numerals can be formed, and the talent which has been expended in the conception of the various plain and ornamental letters is wonderful."—*Standard.*

MEDIÆVAL ALPHABETS AND INITIALS FOR ILLUMINATORS. By F. G. DELAMOTTE. Containing 21 Plates and Illuminated Title, printed in Gold and Colours. With an Introduction by J. WILLIS BROOKS. Fourth and cheaper edition. Small 4to, 4s. ornamental boards.

" A volume in which the letters of the alphabet come forth glorified in gilding and all the colours of the prism interwoven and intermingled."—*Sun.*

THE EMBROIDERER'S BOOK OF DESIGN. Containing Initials, Emblems, Cyphers, Monograms, Ornamental Borders, Ecclesiastical Devices, Mediæval and Modern Alphabets, and National Emblems. Collected by F. DELAMOTTE, and printed in Colours. Oblong royal 8vo, 1s. 6d., ornamental wrapper.

"The book will be of great assistance to ladies and young children who are endowed with the art of plying the needle in this most ornamental and useful pretty work."—*East Anglian Times.*

Wood Carving.

INSTRUCTIONS IN WOOD-CARVING, *for Amateurs;* with Hints on Design. By A LADY. With Ten large Plates, 2s. 6d. in emblematic wrapper.

"The handicraft of the wood-carver, so well as a book can impart it, may be learnt from ' A Lady's' publication."—*Athenæum.*

" The directions given are plain and easily understood."—*English Mechanic.*

Glass Painting.

GLASS STAINING AND THE ART OF PAINTING ON GLASS. From the German of Dr. GESSERT and EMANUEL OTTO FROMBERG. With an Appendix on THE ART OF ENAMELLING. 12mo, 2s. 6d. cloth limp.

Letter Painting.

THE ART OF LETTER PAINTING MADE EASY. By JAMES GREIG BADENOCH. With 12 full-page Engravings of Examples, 1s. cloth limp.

"The system is a simple one, but quite original, and well worth the careful attention of letter-painters. It can be easily mastered and remembered."—*Building News.*

CARPENTRY, TIMBER, etc.

Tredgold's Carpentry, partly Re-written and Enlarged by Tarn.

THE ELEMENTARY PRINCIPLES OF CARPENTRY. A Treatise on the Pressure and Equilibrium of Timber Framing, the Resistance of Timber, and the Construction of Floors, Arches, Bridges, Roofs, Uniting Iron and Stone with Timber, &c. To which is added an Essay on the Nature and Properties of Timber, &c., with Descriptions of the kinds of Wood used in Building; also numerous Tables of the Scantlings of Timber for different purposes, the Specific Gravities of Materials, &c. By THOMAS TREDGOLD, C.E. With an Appendix of Specimens of Various Roofs of Iron and Stone, Illustrated. Seventh Edition, thoroughly revised and considerably enlarged by E. WYNDHAM TARN, M.A., Author of "The Science of Building," &c. With 61 Plates, Portrait of the Author, and several Woodcuts. In one large vol., 4to, price £1 5s. cloth. [*Just published*.

"Ought to be in every architect's and every builder's library."—*Builder*.

"A work whose monumental excellence must commend it wherever skilful carpentry is concerned. The author's principles are rather confirmed than impaired by time. The additional plates are of great intrinsic value."—*Building News*.

Woodworking Machinery.

WOODWORKING MACHINERY: Its Rise, Progress, and Construction. With Hints on the Management of Saw Mills and the Economical Conversion of Timber. Illustrated with Examples of Recent Designs by leading English, French, and American Engineers. By M. POWIS BALE, A.M. Inst. C.E., M.I.M.E. Large crown 8vo, 12s. 6d. cloth.

"Mr. Bale is evidently an expert on the subject, and he has collected so much information that his book is all-sufficient for builders and others engaged in the conversion of timber."—*Architect*.

"The most comprehensive compendium of wood-working machinery we have seen. The author is a thorough master of his subject."—*Building News*.

"The appearance of this book at the present time will, we should think, give a considerable impetus to the onward march of the machinist engaged in the designing and manufacture of wood-working machines. It should be in the office of every wood-working factory."—*English Mechanic*.

Saw Mills.

SAW MILLS: Their Arrangement and Management, and the Economical Conversion of Timber. (Being a Companion Volume to "Woodworking Machinery.") By M. POWIS BALE, A.M. Inst. C.E., M.I.M.E. With numerous Illustrations. Crown 8vo, 10s. 6d. cloth.

"The author is favourably known by his former work on 'Woodworking Machinery,' of which we were able to speak approvingly. This is a companion volume, in which the *administration* of a large sawing establishment is discussed, and the subject examined from a financial standpoint. Hence the size, shape, order, and disposition of saw-mills and the like are gone into in detail, and the course of the timber is traced from its reception to its delivery in its converted state. We could not desire a more complete or practical treatise."—*Builder*.

"We highly recommend Mr. Bale's work to the attention and perusal of all those who are engaged in the art of wood conversion, or who are about building or remodelling saw-mills on improved principles."—*Building News*.

Carpentering.

THE CARPENTER'S NEW GUIDE; or, Book of Lines for Carpenters; comprising all the Elementary Principles essential for acquiring a knowledge of Carpentry. Founded on the late PETER NICHOLSON's Standard Work. A New Edition, revised by ARTHUR ASHPITEL, F.S.A. Together with Practical Rules on Drawing, by GEORGE PYNE. With 74 Plates, 4to, £1 1s. cloth.

Handrailing.

A PRACTICAL TREATISE ON HANDRAILING: Showing New and Simple Methods for Finding the Pitch of the Plank, Drawing the Moulds, Bevelling, Jointing-up, and Squaring the Wreath. By GEORGE COLLINGS. Illustrated with Plates and Diagrams. 12mo, 1s. 6d. cloth limp.

Circular Work.

CIRCULAR WORK IN CARPENTRY AND JOINERY: A Practical Treatise on Circular Work of Single and Double Curvature. By GEORGE COLLINGS, Author of "A Practical Treatise on Handrailing." Illustrated with numerous Diagrams. 12mo, 2s. 6d. cloth limp. [*Just published*.

Timber Merchant's Companion.

THE TIMBER MERCHANT'S AND BUILDER'S COM-PANION. Containing New and Copious Tables of the Reduced Weight and Measurement of Deals and Battens, of all sizes, from One to a Thousand Pieces, and the relative Price that each size bears per Lineal Foot to any given Price per Petersburg Standard Hundred; the Price per Cube Foot of Square Timber to any given Price per Load of 50 Feet; the proportionate Value of Deals and Battens by the Standard, to Square Timber by the Load of 50 Feet; the readiest mode of ascertaining the Price of Scantling per Lineal Foot of any size, to any given Figure per Cube Foot, &c. &c. By WILLIAM DOWSING. Third Edition, Revised and Corrected. Crown 8vo, 3s. cloth.

"Everything is as concise and clear as it can possibly be made. There can be no doubt that every timber merchant and builder ought to possess it."—*Hull Advertiser.*

"An exceedingly well-arranged, clear, and concise manual of tables for the use of all who buy or sell timber."—*Journal of Forestry.*

Practical Timber Merchant.

THE PRACTICAL TIMBER MERCHANT. Being a Guide for the use of Building Contractors, Surveyors, Builders, &c., comprising useful Tables for all purposes connected with the Timber Trade, Marks of Wood, Essay on the Strength of Timber, Remarks on the Growth of Timber, &c. By W. RICHARDSON. Fcap. 8vo, 3s. 6d. cloth.

"This handy manual contains much valuable information for the use of timber merchants, builders, foresters, and all others connected with the growth, sale, and manufacture of timber."—*Journal of Forestry.*

Timber Freight Book.

THE TIMBER MERCHANT'S, SAW MILLER'S, AND IMPORTER'S FREIGHT BOOK AND ASSISTANT. Comprising Rules, Tables, and Memoranda relating to the Timber Trade. By WILLIAM RICHARDSON Timber Broker; together with a Chapter on "Speeds of Saw Mill Machinery," by M. POWIS BALE, M.I.M.E., &c. 12mo, 3s. 6d. cloth boards.

"A very useful manual of rules, tables, and memoranda, relating to the timber trade. We recommend it as a compendium of calculation to all timber measurers and merchants, and as supplying a real want in the trade."—*Building News.*

Tables for Packing-Case Makers.

PACKING-CASE TABLES; showing the number of Superficial Feet in Boxes or Packing-Cases, from six inches square and upwards. By W. RICHARDSON, Timber Broker. Oblong 4to, 3s. 6d. cloth.

"Invaluable labour-saving tables."—*Ironmonger*
"Will save much labour and calculation."—*Grocer.*

Superficial Measurement.

THE TRADESMAN'S GUIDE TO SUPERFICIAL MEASUREMENT. Tables calculated from 1 to 200 inches in length, by 1 to 108 inches in breadth. For the use of Architects, Engineers, Timber Merchants, Builders, &c. By JAMES HAWKINGS. Third Edition. Fcap., 3s. 6d. cloth.

"A useful collection of tables to facilitate rapid calculation of surfaces. The exact area of any surface of which the limits have been ascertained can be instantly determined. The book will be found of the greatest utility to all engaged in building operations."—*Scotsman.*

Forestry.

THE ELEMENTS OF FORESTRY. Designed to afford Information concerning the Planting and Care of Forest Trees for Ornament or Profit, with Suggestions upon the Creation and Care of Woodlands. By F. B. HOUGH. Large crown 8vo, 10s. cloth.

Timber Importer's Guide.

THE TIMBER IMPORTER'S, TIMBER MERCHANT'S AND BUILDER'S STANDARD GUIDE. By RICHARD E. GRANDY. Comprising an Analysis of Deal Standards, Home and Foreign, with Comparative Values and Tabular Arrangements for fixing Nett Landed Cost on Baltic and North American Deals, including all intermediate Expenses, Freight, Insurance, &c. &c. Second Edition, carefully revised. 12mo, 3s. 6d. cloth.

"Everything it pretends to be: built up gradually, it leads one from a forest to a treenail, and throws in, as a makeweight, a host of material concerning bricks, columns, cisterns, &c."—*English Mechanic.*

MINING AND MINING INDUSTRIES.

Metalliferous Mining.

BRITISH MINING : A Treatise on the History, Discovery, Practical Development, and Future Prospects of Metalliferous Mines in the United Kingdom. By ROBERT HUNT, F.R.S., Keeper of Mining Records; Editor of "Ure's Dictionary of Arts, Manufactures, and Mines," &c. Upwards of 950 pp., with 230 Illustrations. Super-royal 8vo, £3 3s. cloth.

∗ OPINIONS OF THE PRESS.

"One of the most valuable works of reference of modern times. Mr. Hunt, as keeper of mining records of the United Kingdom, has had opportunities for such a task not enjoyed by anyone else, and has evidently made the most of them. . . . The language and style adopted are good, and the treatment of the various subjects laborious, conscientious, and scientific."—*Engineering.*

"Probably no one in this country was better qualified than Mr. Hunt for undertaking such a work. Brought into frequent and close association during a long life-time with the principal guardians of our mineral and metallurgical industries, he enjoyed a position exceptionally favourable for collecting the necessary information. The use which he has made of his opportunities is sufficiently attested by the dense mass of information crowded into the handsome volume which has just been published. . . . In placing before the reader a sketch of the present position of British Mining, Mr. Hunt treats his subject so fully and illustrates it so amply that this section really forms a little treatise on practical mining. . . . The book is, in fact, a treasure-house of statistical information on mining subjects, and we know of no other work embodying so great a mass of matter of this kind. Were this the only merit of Mr. Hunt's volume it would be sufficient to render it indispensable in the library of everyone interested in the development of the mining and metallurgical industries of this country."—*Athenæum.*

"A mass of information not elsewhere available, and of the greatest value to those who may be interested in our great mineral industries."—*Engineer.*

"A sound, business-like collection of interesting facts. . . . The amount of information Mr. Hunt has brought together is enormous. . . . The volume appears likely to convey more instruction upon the subject than any work hitherto published."—*Mining Journal.*

"The work will be for the mining industry what Dr. Percy's celebrated treatise has been for the metallurgical—a book that cannot with advantage be omitted from the library."—*Iron and Coal Trades' Review.*

"The literature of mining has hitherto possessed no work approaching in importance to that which has just been published. There is much in Mr. Hunt's valuable work that every shareholder in a mine should read with close attention. The entire subject of practical mining—from the first search for the lode to the latest stages of dressing the ore—is dealt with in a masterly manner." —*Academy.*

Coal and Iron.

THE COAL AND IRON INDUSTRIES OF THE UNITED KINGDOM. Comprising a Description of the Coal Fields, and of the Principal Seams of Coal, with Returns of their Produce and its Distribution, and Analyses of Special Varieties. Also an Account of the occurrence of Iron Ores in Veins or Seams; Analyses of each Variety; and a History of the Rise and Progress of Pig Iron Manufacture since the year 1740, exhibiting the Economies introduced in the Blast Furnaces for its Production and Improvement. By RICHARD MEADE, Assistant Keeper of Mining Records. With Maps of the Coal Fields and Ironstone Deposits of the United Kingdom. 8vo, £1 8s. cloth.

"The book is one which must find a place on the shelves of all interested in coal and iron production, and in the iron, steel, and other metallurgical industries."—*Engineer.*

"Of this book we may unreservedly say that it is the best of its class which we have ever met. . . . A book of reference which no one engaged in the iron or coal trades should omit from his library."—*Iron and Coal Trades' Review.*

"An exhaustive treatise and a valuable work of reference."—*Mining Journal.*

Prospecting.

THE PROSPECTOR'S HANDBOOK: A Guide for the Prospector and Traveller in Search of Metal-Bearing or other Valuable Minerals. By J. W. ANDERSON, M.A. (Camb.), F.R.G.S., Author of "Fiji and New Caledonia." Small crown 8vo, 3s. 6d. cloth. [*Just published.*

"Will supply a much felt want, especially among Colonists, in whose way are so often thrown many mineralogical specimens, the value of which it is difficult for anyone, not a specialist, to determine. The author has placed his instructions before his readers in the plainest possible terms, and his book is the best of its kind."—*Engineer.*

"How to find commercial minerals, and how to identify them when they are found, are the leading points to which attention is directed. The author has managed to pack as much practical detail into his pages as would supply material for a book three times its size."—*Mining Journal.*

"Those toilers who explore the trodden or untrodden tracks on the face of the globe will find much that is useful to them in this book."—*Athenæum.*

Metalliferous Minerals and Mining.

TREATISE ON METALLIFEROUS MINERALS AND MINING. By D. C. DAVIES, F.G.S., Mining Engineer, &c., Author of "A Treatise on Slate and Slate Quarrying." Illustrated with numerous Wood Engravings. Second Edition, carefully Revised. Crown 8vo, 12s. 6d. cloth.

"Neither the practical miner nor the general reader interested in mines, can have a better book for his companion and his guide."—*Mining Journal.*

"The volume is one which no student of mineralogy should be without."—*Colliery Guardian.*

"We are doing our readers a service in calling their attention to this valuable work."—*Mining World.*

"A book that will not only be useful to the geologist, the practical miner, and the metallurgist, but also very interesting to the general public."—*Iron.*

"As a history of the present state of mining throughout the world this book has a real value, and it supplies an actual want, for no such information has hitherto been brought together within such limited space."—*Athenæum.*

Earthy Minerals and Mining.

A TREATISE ON EARTHY AND OTHER MINERALS AND MINING. By D. C. DAVIES, F.G.S. Uniform with, and forming a Companion Volume to, the same Author's "Metalliferous Minerals and Mining." With 76 Wood Engravings. Crown 8vo, 12s. 6d. cloth.

"It is essentially a practical work, intended primarily for the use of practical men.... We do not remember to have met with any English work on mining matters that contains the same amount of information packed in equally convenient form."—*Academy.*

"The book is clearly the result of many years' careful work and thought, and we should be inclined to rank it as among the very best of the handy technical and trades manuals which have recently appeared."—*British Quarterly Review.*

"The subject matter of the volume will be found of high value by all—and they are a numerous class—who trade in earthy minerals."—*Athenæum.*

"Will be found of permanent value for information and reference."—*Iron.*

Underground Pumping Machinery.

MINE DRAINAGE. Being a Complete and Practical Treatise on Direct-Acting Underground Steam Pumping Machinery, with a Description of a large number of the best known Engines, their General Utility and the Special Sphere of their Action, the Mode of their Application, and their merits compared with other forms of Pumping Machinery. By STEPHEN MICHELL. 8vo, 15s. cloth.

"Will be highly esteemed by colliery owners and lessees, mining engineers, and students generally who require to be acquainted with the best means of securing the drainage of mines. It is a most valuable work, and stands almost alone in the literature of steam pumping machinery."—*Colliery Guardian.*

"Much valuable information is given, so that the book is thoroughly worthy of an extensive circulation amongst practical men and purchasers of machinery."—*Mining Journal.*

Mining Tools.

A MANUAL OF MINING TOOLS. For the Use of Mine Managers, Agents, Students, &c. By WILLIAM MORGANS, Lecturer on Practical Mining at the Bristol School of Mines. 12mo, 3s. cloth boards.

ATLAS OF ENGRAVINGS to Illustrate the above, containing 235 Illustrations of Mining Tools, drawn to scale. 4to, 6s. cloth boards.

"Students in the science of mining, and overmen, captains, managers, and viewers may gain practical knowledge and useful hints by the study of Mr. Morgans' manual."—*Colliery Guardian.*

"A valuable work, which will tend materially to improve our mining literature."—*Mining Journal.*

Coal Mining.

COAL AND COAL MINING: A Rudimentary Treatise on. By WARINGTON W. SMYTH, M.A., F.R.S., &c., Chief Inspector of the Mines of the Crown. New Edition, Revised and Corrected. With numerous Illustrations. 12mo, 4s. cloth boards.

"As an outline is given of every known coal-field in this and other countries, as well as of the principal methods of working, the book will doubtless interest a very large number of readers."—*Mining Journal.*

Subterraneous Surveying.

SUBTERRANEOUS SURVEYING, Elementary and Practical Treatise on; with and without the Magnetic Needle. By THOMAS FENWICK, Surveyor of Mines, and THOMAS BAKER, C.E. 12mo, 3s. cloth boards.

NAVAL ARCHITECTURE, NAVIGATION, etc.

Chain Cables.
CHAIN CABLES AND CHAINS. Comprising Sizes and Curves of Links, Studs, &c., Iron for Cables and Chains, Chain Cable and Chain Making, Forming and Welding Links, Strength of Cables and Chains, Certificates for Cables, Marking Cables, Prices of Chain Cables and Chains, Historical Notes, Acts of Parliament, Statutory Tests, Charges for Testing, List of Manufacturers of Cables, &c., &c. By THOMAS W. TRAILL, F.E.R.N., M. Inst. C.E., the Engineer Surveyor in Chief, Board of Trade, the Inspector of Chain Cable and Anchor Proving Establishments, and General Superintendent, Lloyd's Committee on Proving Establishments. With numerous Tables, Illustrations and Lithographic Drawings. Folio, £2 2s. cloth, bevelled boards.

"The author writes not only with a full acquaintance with scientific formulæ and details, but also with a profound and fully-instructed sense of the importance to the safety of our ships and sailors of fidelity in the manufacture of cables. We heartily recommend the book to the specialists to whom it is addressed."—*Athenæum.*

"It contains a vast amount of valuable information. Nothing seems to be wanting to make it a complete and standard work of reference on the subject."—*Nautical Magazine.*

Pocket-Book for Naval Architects and Shipbuilders.
THE NAVAL ARCHITECT'S AND SHIPBUILDER'S POCKET-BOOK of Formulæ, Rules, and Tables, and Marine Engineer's and Surveyor's Handy Book of Reference. By CLEMENT MACKROW, Member of the Institution of Naval Architects, Naval Draughtsman. Third Edition, Revised. With numerous Diagrams, &c. Fcap., 12s. 6d. strongly bound in leather.

"Should be used by all who are engaged in the construction or design of vessels. . . . Will be found to contain the most useful tables and formulæ required by shipbuilders, carefully collected from the best authorities, and put together in a popular and simple form."—*Engineer.*

"The professional shipbuilder has now, in a convenient and accessible form, reliable data for solving many of the numerous problems that present themselves in the course of his work."—*Iron.*

"There is scarcely a subject on which a naval architect or shipbuilder can require to refresh his memory which will not be found within the covers of Mr. Mackrow's book."—*English Mechanic.*

Pocket-Book for Marine Engineers.
A POCKET-BOOK OF USEFUL TABLES AND FORMULÆ FOR MARINE ENGINEERS. By FRANK PROCTOR, A.I.N.A. Third Edition. Royal 32mo, leather, gilt edges, with strap, 4s.

"We recommend it to our readers as going far to supply a long-felt want."—*Naval Science.*
"A most useful companion to all marine engineers."—*United Service Gazette.*

Lighthouses.
EUROPEAN LIGHTHOUSE SYSTEMS. Being a Report of a Tour of Inspection made in 1873. By Major GEORGE H. ELLIOT, Corps of Engineers, U.S.A. With 51 Engravings and 31 Woodcuts. 8vo, 21s. cloth.

** *The following are published in* WEALE'S RUDIMENTARY SERIES.

MASTING, MAST-MAKING, AND RIGGING OF SHIPS. By ROBERT KIPPING, N.A. Fifteenth Edition. 12mo, 2s. 6d. cloth boards.

SAILS AND SAIL-MAKING. Eleventh Edition, Enlarged, with an Appendix. By ROBERT KIPPING, N.A. Illustrated. 12mo, 3s. cloth boards.

NAVAL ARCHITECTURE. By JAMES PEAKE. Fifth Edition with Plates and Diagrams. 12mo, 4s. cloth boards.

MARINE ENGINES AND STEAM VESSELS (*A Treatise on*). By ROBERT MURRAY, C.E., Principal Officer to the Board of Trade for the East Coast of Scotland District. Eighth Edition, thoroughly Revised, with considerable Additions, by the Author and by GEORGE CARLISLE, C.E., Senior Surveyor to the Board of Trade at Liverpool. 12mo, 5s. cloth boards.

PRACTICAL NAVIGATION. Consisting of the Sailor's Sea-Book, by JAMES GREENWOOD and W. H. ROSSER; together with the requisite Mathematical and Nautical Tables for the Working of the Problems, by HENRY LAW, C.E., and Professor J. R. YOUNG. 12mo, 7s., half-bound.

NATURAL PHILOSOPHY AND SCIENCE.

Text Book of Electricity.

THE STUDENT'S TEXT-BOOK OF ELECTRICITY. By HENRY M. NOAD, Ph.D., F.R.S., F.C.S. New Edition, carefully Revised. With an Introduction and Additional Chapters, by W. H. PREECE, M.I.C.E., Vice-President of the Society of Telegraph Engineers, &c. With 470 Illustrations. Crown 8vo, 12s. 6d. cloth.

"The original plan of this book has been carefully adhered to so as to make it a reflex of the existing state of electrical science, adapted for students. . . . Discovery seems to have progressed with marvellous strides; nevertheless it has now apparently ceased, and practical applications have commenced their career; and it is to give a faithful account of these that this fresh edition of Dr. Noad's valuable text-book is launched forth."—*Extract from Introduction by W. H. Preece, Esq.*

"We can recommend Dr. Noad's book for clear style, great range of subject, a good index, and a plethora of woodcuts. Such collections as the present are indispensable."—*Athenæum.*

"An admirable text-book for every student—beginner or advanced—of electricity."—*Engineering.*

"Dr. Noad's text-book has earned for itself the reputation of a truly scientific manual for the student of electricity, and we gladly hail this new amended edition, which brings it once more to the front. Mr. Preece as reviser, with the assistance of Mr. H. R. Kempe and Mr. J. P. Edwards, has added all the practical results of recent invention and research to the admirable theoretical expositions of the author, so that the book is about as complete and advanced as it is possible for any book to be within the limits of a text-book."—*Telegraphic Journal.*

Electricity.

A MANUAL OF ELECTRICITY: Including Galvanism, Magnetism, Dia-Magnetism, Electro-Dynamics, Magno-Electricity, and the Electric Telegraph. By HENRY M. NOAD, Ph.D., F.R.S., F.C.S. Fourth Edition. With 500 Woodcuts. 8vo, £1 4s. cloth.

"The accounts given of electricity and galvanism are not only complete in a scientific sense, but, which is a rarer thing, are popular and interesting."—*Lancet.*

"It is worthy of a place in the library of every public institution."—*Mining Journal.*

Electric Light.

ELECTRIC LIGHT: Its Production and Use. Embodying Plain Directions for the Treatment of Voltaic Batteries, Electric Lamps, and Dynamo-Electric Machines. By J. W. URQUHART, C.E., Author of "Electroplating: A Practical Handbook." Edited by F. C. WEBB, M.I.C.E., M.S.T.E. Second Edition, Revised, with large Additions and 128 Illusts. 7s. 6d. cloth.

"The book is by far the best that we have yet met with on the subject."—*Athenæum.*

"It is the only work at present available which gives, in language intelligible for the most part to the ordinary reader, a general but concise history of the means which have been adopted up to the present time in producing the electric light."—*Metropolitan*

Electric Lighting.

THE ELEMENTARY PRINCIPLES OF ELECTRIC LIGHTING. By ALAN A. CAMPBELL SWINTON, Associate S.T.E. Crown 8vo, 1s. 6d., cloth. [*Just published.*

"As a stepping-stone to treatises of a more advanced nature, this little work will be found most efficient."—*Bookseller.*

"Anyone who desires a short and thoroughly clear exposition of the elementary principles of electric-lighting cannot do better than read this little work."—*Bradford Observer.*

Dr. Lardner's School Handbooks.

NATURAL PHILOSOPHY FOR SCHOOLS. By Dr. LARDNER. 328 Illustrations. Sixth Edition. One Vol., 3s. 6d. cloth.

"A very convenient class-book for junior students in private schools. It is intended to convey, in clear and precise terms, general notions of all the principal divisions of Physical Science."—*British Quarterly Review.*

ANIMAL PHYSIOLOGY FOR SCHOOLS. By Dr. LARDNER. With 190 Illustrations. Second Edition. One Vol., 3s. 6d. cloth.

"Clearly written, well arranged, and excellently illustrated."—*Gardner's Chronicle.*

Dr. Lardner's Electric Telegraph.

THE ELECTRIC TELEGRAPH. By Dr. LARDNER. Revised and Re-written by E. B. BRIGHT, F.R.A.S. 140 Illustrations. Small 8vo, 2s. 6d. cloth.

"One of the most readable books extant on the Electric Telegraph."—*English Mechanic.*

NATURAL PHILOSOPHY AND SCIENCE. 21

Storms.

STORMS: Their Nature, Classification, and Laws; with the Means of Predicting them by their Embodiments, the Clouds. By WILLIAM BLASIUS. With Coloured Plates and numerous Wood Engravings. Crown 8vo, 10s. 6d. cloth.

"A useful repository to meteorologists in the study of atmospherical disturbances. Will repay perusal as being the production of one who gives evidence of acute observation."—*Nature.*

The Blowpipe.

THE BLOWPIPE IN CHEMISTRY, MINERALOGY, AND GEOLOGY. Containing all known Methods of Anhydrous Analysis, many Working Examples, and Instructions for Making Apparatus. By Lieut.-Colonel W. A. ROSS, R.A., F.G.S. With 120 Illustrations. Crown 8vo, 3s. 6d. cloth.

"The student who goes conscientiously through the course of experimentation here laid down will gain a better insight into inorganic chemistry and mineralogy than if he had 'got up' any of the best text-books of the day, and passed any number of examinations."—*Chemical News.*

The Military Sciences.

AIDE-MEMOIRE TO THE MILITARY SCIENCES. Framed from Contributions of Officers and others connected with the different Services. Originally edited by a Committee of the Corps of Royal Engineers. Second Edition, most carefully revised by an Officer of the Corps, with many Additions; containing nearly 350 Engravings and many hundred Woodcuts. Three Vols., royal 8vo, extra cloth boards, and lettered, £4 10s.

"A compendious encyclopædia of military knowledge, to which we are greatly indebted."—*Edinburgh Review.*

"The most comprehensive work of reference to the military and collateral sciences.'—*Volunteer Service Gazette.*

Field Fortification.

A TREATISE ON FIELD FORTIFICATION, THE ATTACK OF FORTRESSES, MILITARY MINING, AND RECONNOITRING. By Colonel I. S. MACAULAY, late Professor of Fortification in the R.M.A., Woolwich. Sixth Edition, crown 8vo, cloth, with separate Atlas of 12 Plates, 12s.

Conchology.

MANUAL OF THE MOLLUSCA: A Treatise on Recent and Fossil Shells. By Dr. S. P. WOODWARD, A.L.S. With Appendix by RALPH TATE, A.L.S., F.G.S. With numerous Plates and 300 Woodcuts. Cloth boards, 7s. 6d.

"A most valuable storehouse of conchological and geological information."—*Hardwicke's Science Gossip.*

Astronomy.

ASTRONOMY. By the late Rev. ROBERT MAIN, M.A., F.R.S., formerly Radcliffe Observer at Oxford. Third Edition, Revised and Corrected to the present time, by WILLIAM THYNNE LYNN, B.A., F.R.A.S., formerly of the Royal Observatory, Greenwich. 12mo, 2s. cloth limp.

"A sound and simple treatise, carefully edited, and a capital book for beginners."—*Knowledge·*
"Accurately brought down to the requirements of the present time."—*Educational Times*

Geology.

RUDIMENTARY TREATISE ON GEOLOGY, PHYSICAL AND HISTORICAL. Consisting of "Physical Geology," which sets forth the leading Principles of the Science; and "Historical Geology," which treats of the Mineral and Organic Conditions of the Earth at each successive epoch, especial reference being made to the British Series of Rocks. By RALPH TATE, A.L.S., F.G.S., &c., &c. With 250 Illustrations. 12mo, 5s. cloth boards.

"The fulness of the matter has elevated the book into a manual. Its Information is exhaustive and well arranged."—*School Board Chronicle.*

Geology and Genesis.

THE TWIN RECORDS OF CREATION; or, Geology and Genesis: their Perfect Harmony and Wonderful Concord. By GEORGE W. VICTOR LE VAUX. Numerous Illustrations. Fcap. 8vo, 5s. cloth.

"A valuable contribution to the evidences of revelation, and disposes very conclusively of the arguments of those who would set God's Works against God's Word. No real difficulty is shirked, and no sophistry is left unexposed."—*The Rock.*

Dr. LARDNER'S HANDBOOKS of NATURAL PHILOSOPHY.

** The following five volumes, though each is complete in itself, and to be purchased separately, form A COMPLETE COURSE OF NATURAL PHILOSOPHY. The style is studiously popular. It has been the author's aim to supply Manuals for the Student, the Engineer, the Artisan, and the superior classes in Schools.

THE HANDBOOK OF MECHANICS. Enlarged and almost rewritten by BENJAMIN LOEWY, F.R.A.S. With 378 Illustrations. Post 8vo, 6s. cloth.

"The perspicuity of the original has been retained, and chapters which had become obsolete have been replaced by others of more modern character. The explanations throughout are studiously popular, and care has been taken to show the application of the various branches of physics to the industrial arts, and to the practical business of life."—*Mining Journal.*

"Mr. Loewy has carefully revised the book, and brought it up to modern requirements."—*Nature.*

"Natural philosophy has had few exponents more able or better skilled in the art of popularising the subject than Dr. Lardner; and Mr. Loewy is doing good service in fitting this treatise, and the others of the series, for use at the present time."—*Scotsman.*

THE HANDBOOK OF HYDROSTATICS AND PNEUMATICS. New Edition, Revised and Enlarged, by BENJAMIN LOEWY, F.R.A.S. With 236 Illustrations. Post 8vo, 5s. cloth.

"For those 'who desire to attain an accurate knowledge of physical science without the profound methods of mathematical investigation,' this work is not merely intended, but well adapted."—*Chemical News.*

"The volume before us has been carefully edited, augmented to nearly twice the bulk of the former edition, and all the most recent matter has been added. . . . It is a valuable text-book."—*Nature.*

"Candidates for pass examinations will find it, we think, specially suited to their requirements." *English Mechanic.*

THE HANDBOOK OF HEAT. Edited and almost entirely rewritten by BENJAMIN LOEWY, F.R.A.S., &c. 117 Illustrations. Post 8vo, 6s. cloth.

"The style is always clear and precise, and conveys instruction without leaving any cloudiness or lurking doubts behind."—*Engineering.*

"A most exhaustive book on the subject on which it treats, and is so arranged that it can be understood by all who desire to attain an accurate knowledge of physical science. . . . . Mr. Loewy has included all the latest discoveries in the varied laws and effects of heat."—*Standard.*

"A complete and handy text-book for the use of students and general readers."—*English Mechanic.*

THE HANDBOOK OF OPTICS. By DIONYSIUS LARDNER, D.C.L. formerly Professor of Natural Philosophy and Astronomy in University College, London. New Edition. Edited by T. OLVER HARDING, B.A. Lond., of University College, London. With 298 Illustrations. Small 8vo, 448 pages, 5s. cloth.

"Written by one of the ablest English scientific writers, beautifully and elaborately illustrated."—*Mechanics' Magazine.*

THE HANDBOOK OF ELECTRICITY, MAGNETISM, AND ACOUSTICS. By Dr. LARDNER. Ninth Thousand. Edit. by GEORGE CAREY FOSTER, B.A., F.C.S. With 400 Illustrations. Small 8vo, 5s. cloth.

"The book could not have been entrusted to anyone better calculated to preserve the terse and lucid style of Lardner, while correcting his errors and bringing up his work to the present state of scientific knowledge."—*Popular Science Review.*

Dr. Lardner's Handbook of Astronomy.

THE HANDBOOK OF ASTRONOMY. Forming a Companion to the "Handbook of Natural Philosophy." By DIONYSIUS LARDNER, D.C.L., formerly Professor of Natural Philosophy and Astronomy in University College, London. Fourth Edition. Revised and Edited by EDWIN DUNKIN, F.R.A.S., Royal Observatory, Greenwich. With 38 Plates and upwards of 100 Woodcuts. In One Vol., small 8vo, 550 pages, 9s. 6d. cloth.

"Probably no other book contains the same amount of information in so compendious and well-arranged a form—certainly none at the price at which this is offered to the public.'—*Athenæum.*

"We can do no other than pronounce this work a most valuable manual of astronomy, and we strongly recommend it to all who wish to acquire a general—but at the same time correct—acquaintance with this sublime science."—*Quarterly Journal of Science.*

"One of the most deservedly popular books on the subject . . . We would recommend not only the student of the elementary principles of the science, but he who aims at mastering the higher and mathematical branches of astronomy, not to be without this work beside him."—*Practical Magazine.*

DR. LARDNER'S MUSEUM OF SCIENCE AND ART.

THE MUSEUM OF SCIENCE AND ART. Edited by DIONYSIUS LARDNER, D.C.L., formerly Professor of Natural Philosophy and Astronomy in University College, London. With upwards of 1,200 Engravings on Wood. In 6 Double Volumes, £1 1s., in a new and elegant cloth binding; or handsomely bound in half-morocco, 31s. 6d.

Contents:

The Planets: Are they Inhabited Worlds?—Weather Prognostics—Popular Fallacies in Questions of Physical Science—Latitudes and Longitudes—Lunar Influences—Meteoric Stones and Shooting Stars—Railway Accidents—Light—Common Things: Air—Locomotion in the United States—Cometary Influences—Common Things: Water—The Potter's Art—Common Things: Fire—Locomotion and Transport, their Influence and Progress—The Moon—Common Things: The Earth—The Electric Telegraph—Terrestrial Heat—The Sun—Earthquakes and Volcanoes—Barometer, Safety Lamp, and Whitworth's Micrometric Apparatus—Steam—The Steam Engine—The Eye—The Atmosphere—Time—Common Things: Pumps—Common Things: Spectacles, the Kaleidoscope—Clocks and Watches—Microscopic Drawing and Engraving—Locomotive—Thermometer—New Planets: Leverrier and Adams's Planet—Magnitude and Minuteness—Common Things: The Almanack—Optical Images—How to observe the Heavens—Common Things: The Looking-glass—Stellar Universe—The Tides—Colour—Common Things: Man—Magnifying Glasses—Instinct and Intelligence—The Solar Microscope—The Camera Lucida—The Magic Lantern—The Camera Obscura—The Microscope—The White Ants: Their Manners and Habits—The Surface of the Earth, or First Notions of Geography—Science and Poetry—The Bee—Steam Navigation—Electro-Motive Power—Thunder, Lightning, and the Aurora Borealis—The Printing Press—The Crust of the Earth—Comets—The Stereoscope—The Pre-Adamite Earth—Eclipses—Sound.

Opinions of the Press.

"This series, besides affording popular but sound instruction on scientific subjects, with which the humblest man in the country ought to be acquainted, also undertakes that teaching of 'Common Things' which every well-wisher of his kind is anxious to promote. Many thousand copies of this serviceable publication have been printed, in the belief and hope that the desire for instruction and improvement widely prevails; and we have no fear that such enlightened faith will meet with disappointment."—*Times.*

"A cheap and interesting publication, alike informing and attractive. The papers combine subjects of importance and great scientific knowledge, considerable inductive powers, and a popular style of treatment."—*Spectator.*

"The 'Museum of Science and Art' is the most valuable contribution that has ever been made to the Scientific Instruction of every class of society."—Sir DAVID BREWSTER, in the *North British Review.*

"Whether we consider the liberality and beauty of the illustrations, the charm of the writing, or the durable interest of the matter, we must express our belief that there is hardly to be found among the new books one that would be welcomed by people of so many ages and classes as a valuable present."—*Examiner.*

*** *Separate books formed from the above, suitable for Workmen's Libraries, Science Classes, &c.*

Common Things Explained. Containing Air, Earth, Fire, Water, Time, Man, the Eye, Locomotion, Colour, Clocks and Watches, &c. 233 Illustrations, cloth gilt, 5s.

The Microscope. Containing Optical Images, Magnifying Glasses, Origin and Description of the Microscope, Microscopic Objects, the Solar Microscope, Microscopic Drawing and Engraving, &c. 147 Illustrations, cloth gilt, 2s.

Popular Geology. Containing Earthquakes and Volcanoes, the Crust of the Earth, &c. 201 Illustrations, cloth gilt, 2s. 6d.

Popular Physics. Containing Magnitude and Minuteness, the Atmosphere, Meteoric Stones, Popular Fallacies, Weather Prognostics, the Thermometer, the Barometer, Sound, &c. 85 Illustrations, cloth gilt, 2s. 6d.

Steam and its Uses. Including the Steam Engine, the Locomotive, and Steam Navigation. 89 Illustrations, cloth gilt, 2s.

Popular Astronomy. Containing How to observe the Heavens—The Earth, Sun, Moon, Planets, Light, Comets, Eclipses, Astronomical Influences, &c. 182 Illustrations, 4s. 6d.

The Bee and White Ants: Their Manners and Habits. With Illustrations of Animal Instinct and Intelligence. 135 Illustrations, cloth gilt, 2s.

The Electric Telegraph Popularised. To render intelligible to all who can Read, irrespective of any previous Scientific Acquirements, the various forms of Telegraphy in Actual Operation. 100 Illustrations, cloth gilt, 1s. 6d.

MATHEMATICS, GEOMETRY, TABLES, etc.

Practical Mathematics.

MATHEMATICS FOR PRACTICAL MEN. Being a Common-place Book of Pure and Mixed Mathematics. Designed chiefly for the Use of Civil Engineers, Architects and Surveyors. By OLINTHUS GREGORY, LL.D., F.R.A.S., Enlarged by HENRY LAW, C.E. 4th Edition, carefully Revised by J. R. YOUNG, formerly Professor of Mathematics, Belfast College. With 13 Plates, 8vo, £1 1s. cloth.

"The engineer or architect will here find ready to his hand rules for solving nearly every mathematical difficulty that may arise in his practice. The rules are in all cases explained by means of examples, in which every step of the process is clearly worked out."—*Builder.*

"One of the most serviceable books for practical mechanics. . . . It is an instructive book for the student, and a Text-book for him who, having once mastered the subjects it treats of, needs occasionally to refresh his memory upon them."—*Building News.*

Metrical Units and Systems, etc.

MODERN METROLOGY: A Manual of the Metrical Units and Systems of the Present Century. With an Appendix containing a proposed English System. By LOWIS D'A. JACKSON, A.M. Inst. C.E., Author of "Aid to Survey Practice," &c. Large crown 8vo, 12s. 6d. cloth.

"The author has brought together much valuable and interesting information. . . . We cannot but recommend the work to the consideration of all interested in the practical reform of our weights and measures."—*Nature.*

"For exhaustive tables of equivalent weights and measures of all sorts, and for clear demonstrations of the effects of the various systems that have been proposed or adopted, Mr. Jackson's treatise is without a rival."—*Academy.*

The Metric System.

A SERIES OF METRIC TABLES, in which the British Standard Measures and Weights are compared with those of the Metric System at present in Use on the Continent. By C. H. DOWLING, C.E. 8vo, 10s. 6d. strongly bound.

"Their accuracy has been certified by Professor Airy, the Astronomer-Royal."—*Builder.*

"Mr. Dowling's Tables are well put together as a ready-reckoner for the conversion of one system into the other."—*Athenæum*

Geometry for the Architect, Engineer, etc.

PRACTICAL GEOMETRY, for the Architect, Engineer and Mechanic. Giving Rules for the Delineation and Application of various Geometrical Lines, Figures and Curves. By E. W. TARN, M.A., Architect, Author of "The Science of Building," &c. Second Edition. With Appendices on Diagrams of Strains and Isometrical Projection. With 172 Illustrations, demy 8vo, 9s. cloth.

"No book with the same objects in view has ever been published in which the clearness of the rules laid down and the illustrative diagrams have been so satisfactory."—*Scotsman.*

"This is a manual for the practical man, whether architect, engineer, or mechanic. . . . The object of the author being to avoid all abstruse formulæ or complicated methods, and to enable persons with but a moderate knowledge of geometry to work out the problems required."—*English Mechanic.*

The Science of Geometry.

THE GEOMETRY OF COMPASSES; or, Problems Resolved by the mere Description of Circles and the use of Coloured Diagrams and Symbols. By OLIVER BYRNE. Coloured Plates. Crown 8vo, 3s. 6d. cloth.

"The treatise is a good one, and remarkable—like all Mr. Byrne's contributions to the science of geometry—for the lucid character of its teaching."—*Building News.*

Iron and Metal Trades' Calculator.

THE IRON AND METAL TRADES' COMPANION. For expeditiously ascertaining the Value of any Goods bought or sold by Weight, from 1s. per cwt. to 112s. per cwt., and from one farthing per pound to one shilling per pound. Each Table extends from one pound to 100 tons. To which are appended Rules on Decimals, Square and Cube Root, Mensuration of Superficies and Solids, &c.; Tables of Weights of Materials, and other Useful Memoranda. By THOS. DOWNIE. 396 pp., 9s. Strongly bound leather.

"A most useful set of tables, and will supply a want, for nothing like them before existed."—*Building News.*

"Although specially adapted to the iron and metal trades, the tables will be found useful in every other business in which merchandise is bought and sold by weight."—*Railway News.*

MATHEMATICS, GEOMETRY, TABLES, etc.

Calculator for Numbers and Weights Combined.

THE COMBINED NUMBER AND WEIGHT CALCULATOR. Containing upwards of 250,000 Separate Calculations, showing at a glance the value at 421 different rates, ranging from $\frac{1}{16}$th of a Penny to 20s. each, or per cwt., and £20 per ton, of any number of articles consecutively, from 1 to 470.—Any number of cwts., qrs., and lbs., from 1 cwt. to 470 cwts.—Any number of tons, cwts., qrs., and lbs., from 1 to 23½ tons. By WILLIAM CHADWICK, Public Accountant. Imp. 8vo, 30s., strongly bound.

☞ *This comprehensive and entirely unique and original Calculator is adapted for the use of Accountants and Auditors, Railway Companies, Canal Companies, Shippers, Shipping Agents, General Carriers, &c.*
Ironfounders, Brassfounders, Metal Merchants, Iron Manufacturers, Ironmongers, Engineers, Machinists, Boiler Makers, Millwrights, Roofing, Bridge and Girder Makers, Colliery Proprietors, &c.
Timber Merchants, Builders, Contractors, Architects, Surveyors, Auctioneers, Valuers, Brokers, Mill Owners and Manufacturers, Mill Furnishers, Merchants and General Wholesale Tradesmen.

*** OPINIONS OF THE PRESS.

"The book contains the answers to questions, and not simply a set of ingenious puzzle methods of arriving at results. It is as easy of reference for any answer or any number of answers as a dictionary, and the references are even more quickly made. For making up accounts or estimates, the book must prove invaluable to all who have any considerable quantity of calculations involving price and measure in any combination to do."—*Engineer.*

"The most complete and practical ready reckoner which it has been our fortune yet to see. It is difficult to imagine a trade or occupation in which it could not be of the greatest use, either in saving human labour or in checking work. The publishers have placed within the reach of every commercial man an invaluable and unfailing assistant."—*The Miller.*

Comprehensive Weight Calculator.

THE WEIGHT CALCULATOR. Being a Series of Tables upon a New and Comprehensive Plan, exhibiting at One Reference the exact Value of any Weight from 1 lb. to 15 tons, at 300 Progressive Rates, from 1d. to 168s. per cwt., and containing 186,000 Direct Answers, which, with their Combinations, consisting of a single addition (mostly to be performed at sight), will afford an aggregate of 10,266,000 Answers; the whole being calculated and designed to ensure correctness and promote despatch. By HENRY HARBEN, Accountant. An entirely New Edition, carefully Revised. Royal 8vo, strongly half-bound, £1 5s.

"Of priceless value to business men. Its accuracy and completeness have secured for it a reputation which renders it quite unnecessary for us to say one word in its praise. It is a necessary book in all mercantile offices."—*Sheffield Independent.*

Comprehensive Discount Guide.

THE DISCOUNT GUIDE. Comprising several Series of Tables for the use of Merchants, Manufacturers, Ironmongers, and others, by which may be ascertained the exact Profit arising from any mode of using Discounts, either in the Purchase or Sale of Goods, and the method of either Altering a Rate of Discount or Advancing a Price, so as to produce, by one operation, a sum that will realise any required profit after allowing one or more Discounts: to which are added Tables of Profit or Advance from 1¼ to 90 per cent., Tables of Discount from 1¼ to 98¾ per cent., and Tables of Commission, &c., from ⅛ to 10 per cent. By HENRY HARBEN, Accountant, Author of "The Weight Calculator." New Edition, carefully Revised and Corrected. Demy 8vo, 544 pp. half-bound, £1 5s.

"A book such as this can only be appreciated by business men, to whom the saving of time means saving of money. We have the high authority of Professor J. R. Young that the tables throughout the work are constructed upon strictly accurate principles. The work must prove of great value to merchants, manufacturers, and general traders."—*British Trade Journal*

Iron Shipbuilders' and Iron Merchants' Tables.

IRON-PLATE WEIGHT TABLES: For Iron Shipbuilders, Engineers and Iron Merchants. Containing the Calculated Weights of upwards of 150,000 different sizes of Iron Plates, from 1 foot by 6 in. by ¼ in. to 10 feet by 5 feet by 1 in. Worked out on the basis of 40 lbs. to the square foot of Iron of 1 inch in thickness. Carefully compiled and thoroughly Revised by H. BURLINSON and W. H. SIMPSON. Oblong 4to, 25s. half-bound.

"This work will be found of great utility. The authors have had much practical experience of what is wanting in making estimates; and the use of the book will save much time in making elaborate calculations."—*English Mechanic.*

INDUSTRIAL AND USEFUL ARTS.

Soap-making.

THE ART OF SOAP-MAKING: A Practical Handbook of the Manufacture of Hard and Soft Soaps, Toilet Soaps, &c. Including many New Processes, and a Chapter on the Recovery of Glycerine from Waste Leys. By ALEXANDER WATT, Author of "Electro-Metallurgy Practically Treated," &c. With numerous Illustrations. Second Edition, Revised. Crown 8vo, 9s. cloth.

"The work will prove very useful, not merely to the technological student, but to the practical soapboiler who wishes to understand the theory of his art."—*Chemical News.*

"It is really an excellent example of a technical manual, entering, as it does, thoroughly and exhaustively both into the theory and practice of soap manufacture."—*Knowledge.*

"Mr. Watt's book is a thoroughly practical treatise on an art which has almost no literature in our language. We congratulate the author on the success of his endeavour to fill a void in English technical literature."—*Nature.*

Leather Manufacture.

THE ART OF LEATHER MANUFACTURE. Being a Practical Handbook, in which the Operations of Tanning, Currying, and Leather Dressing are fully Described, and the Principles of Tanning Explained, and many Recent Processes introduced; as also Methods for the Estimation of Tannin, and a Description of the Arts of Glue Boiling, Gut Dressing, &c. By ALEXANDER WATT, Author of "Soap-Making," "Electro-Metallurgy," &c. With numerous Illustrations. Crown 8vo, 12s. 6d. cloth.

"Mr. Watt has rendered an important service to the trade, and no less to the student of technology."—*Chemical News.*

"A sound, comprehensive treatise. The book is an eminently valuable production which redounds to the credit of both author and publishers."—*Chemical Review.*

"This volume is technical without being tedious, comprehensive and complete without being prosy, and it bears on every page the impress of a master hand. We have never come across a better trade treatise, nor one that so thoroughly supplied an absolute want."—*Shoe and Leather Trades' Chronicle.*

Boot and Shoe Making.

THE ART OF BOOT AND SHOE-MAKING. A Practical Handbook, including Measurement, Last-Fitting, Cutting-Out, Closing and Making, with a Description of the most approved Machinery employed. By JOHN B. LENO, late Editor of *St. Crispin*, and *The Boot and Shoe-Maker*. With numerous Illustrations. Crown 8vo, 5s. cloth.

"This excellent treatise is by far the best work ever written on the subject. A new work, embracing all modern improvements, was much wanted. This want is now satisfied. The chapter on clicking, which shows how waste may be prevented, will save fifty times the price of the book."—*Scottish Leather Trader.*

Dentistry.

MECHANICAL DENTISTRY: A Practical Treatise on the Construction of the various kinds of Artificial Dentures. Comprising also Useful Formulæ, Tables and Receipts for Gold Plate, Clasps, Solders, &c. &c. By CHARLES HUNTER. Second Edition, Revised. With upwards of 100 Wood Engravings. Crown 8vo, 7s. 6d. cloth.

"We can strongly recommend Mr. Hunter's treatise to all students preparing for the profession of dentistry, as well as to every mechanical dentist."—*Dublin Journal of Medical Science.*

"A work in a concise form that few could read without gaining information from."—*British Journal of Dental Science.*

Brewing.

A HANDBOOK FOR YOUNG BREWERS. By HERBERT EDWARDS WRIGHT, B.A. Crown 8vo, 3s. 6d. cloth.

"This little volume, containing such a large amount of good sense in so small a compass, ought to recommend itself to every brewery pupil, and many who have passed that stage."—*Brewers' Guardian.*

"The book is very clearly written, and the author has successfully brought his scientific knowledge to bear upon the various processes and details of brewing."—*Brewer.*

Wood Engraving.

A PRACTICAL MANUAL OF WOOD ENGRAVING. With a Brief Account of the History of the Art. By WILLIAM NORMAN BROWN. With numerous Illustrations. Crown 8vo, 2s. cloth.

"The author deals with the subject in a thoroughly practical and easy series of representative lessons."—*Paper and Printing Trades' Journal.*

INDUSTRIAL AND USEFUL ARTS. 27

Electrolysis of Gold, Silver, Copper, &c.

ELECTRO-DEPOSITION : *A Practical Treatise on the Electrolysis of Gold, Silver, Copper, Nickel, and other Metals and Alloys.* With descriptions of Voltaic Batteries, Magnets and Dynamo-Electric Machines, Thermopiles, and of the Materials and Processes used in every Department of the Art, and several Chapters on ELECTRO-METALLURGY. By ALEXANDER WATT, Author of "Electro-Metallurgy," "The Art of Soapmaking." &c. With numerous Illustrations. Crown 8vo, 12s. 6d., cloth.

"Evidently written by a practical man who has spent a long period of time in electro-plate workshops. The information given respecting the details of workshop manipulation is remarkably complete. . . . Mr. Watt's book will prove of great value to electro-depositors, jewellers, and various other workers in metal."—*Nature.*

"Eminently a book for the practical worker in electro-deposition. It contains minute and practical descriptions of methods, processes and materials as actually pursued and used in the workshop. Mr. Watt's book recommends itself to all interested in its subjects. —*Engineer.*

"Contains an enormous quantity of practical information; and there are probably few items omitted which could be of any possible utility to workers in galvano-plasty. As a practical manual the book can be recommended to all who wish to study the art of electro-deposition."—*English Mechanic.*

Electroplating, etc.

ELECTROPLATING : *A Practical Handbook.* By J. W. URQUHART, C.E. With numerous Illustrations. Crown 8vo, 5s. cloth.

"The information given appears to be based on direct personal knowledge. . . Its science is sound and the style is always clear."—*Athenæum.*

Electrotyping, etc.

ELECTROTYPING : *The Reproduction and Multiplication of Printing Surfaces and Works of Art by the Electro-deposition of Metals.* By J. W. URQUHART, C.E. Crown 8vo, 5s. cloth.

"The book is thoroughly practical. The reader is, therefore, conducted through the leading laws of electricity, then through the metals used by electrotypers, the apparatus, and the depositing processes, up to the final preparation of the work."—*Art Journal.*

"We can recommend this treatise, not merely to amateurs, but to those actually engaged in the trade."—*Chemical News.*

Electro-Metallurgy.

ELECTRO-METALLURGY; *Practically Treated.* By ALEXANDER WATT, F.R.S.S.A. Eighth Edition, Revised, with Additional Matter and Illustrations, including the most recent Processes. 12mo, 3s. 6d. cloth boards.

"From this book both amateur and artisan may learn everything necessary for the successful prosecution of electroplating."—*Iron.*

Goldsmiths' Work.

THE GOLDSMITH'S HANDBOOK. By GEORGE E. GEE, Jeweller, &c. Third Edition, considerably Enlarged. 12mo, 3s. 6d. cloth boards.

"A good, sound, technical educator, and will be generally accepted as an authority. It is essentially a book for the workshop, and exactly fulfils the purpose intended."—*Horological Journal.*

"Will speedily become a standard book which few will care to be without."—*Jeweller and Metalworker.*

Silversmiths' Work.

THE SILVERSMITH'S HANDBOOK. By GEORGE E. GEE, Jeweller, &c. Second Edition, Revised, with numerous Illustrations. 12mo 3s. 6d. cloth boards.

"The chief merit of the work is its practical character. . . The workers in the trade will speedily discover its merits when they sit down to study it."—*English Mechanic.*

"This work forms a valuable sequel to the author's 'Goldsmith's Handbook.'"—*Silversmiths' Trade Journal.*

*** *The above two works together, strongly half-bound, price 7s.*

Textile Manufacturers' Tables.

UNIVERSAL TABLES OF TEXTILE STRUCTURE. For the use of Manufacturers in every branch of Textile Trade. By JOSEPH EDMONDSON. Oblong folio, strongly bound in cloth, price 7s. 6d.

☞ *These Tables provide what has long been wanted, a simple and easy means of adjusting yarns to "reeds" or "setts," or to "picks" or "shots," and vice versa, so that fabrics may be made of varying weights or fineness, but having the same character and proportions.*

CHEMICAL MANUFACTURES & COMMERCE.

The Alkali Trade, Manufacture of Sulphuric Acid, etc.

A MANUAL OF THE ALKALI TRADE, including the Manufacture of Sulphuric Acid, Sulphate of Soda, and Bleaching Powder. By JOHN LOMAS, Alkali Manufacturer, Newcastle-upon-Tyne and London. With 232 Illustrations and Working Drawings, and containing 390 pages of Text. Second Edition, with Additions. Super-royal 8vo, £1 10s. cloth.

⁎⁎* This work provides (1) a Complete Handbook for intending Alkali and Sulphuric Acid Manufacturers, and for those already in the field who desire to improve their plant, or to become practically acquainted with the latest processes and developments of the trade: (2) a Handy Volume which Manufacturers can put into the hands of their Managers and Foremen as a useful guide in their daily rounds of duty.

"The author has given the fullest, most practical, and, to all concerned in the alkali trade, most valuable mass of information that, to our knowledge, has been published."—*Engineer.*

"This book is written by a manufacturer for manufacturers. The working details of the most approved forms of apparatus are given, and these are accompanied by no less than 232 wood engravings, all of which may be used for the purposes of construction. Every step in the manufacture is very fully described in this manual, and each improvement explained. Everything which tends to introduce economy into the technical details of this trade receives the fullest attention."—*Athenæum.*

"The author is not one of those clever compilers who, on short notice, will 'read up' any conceivable subject, but a practical man in the best sense of the word. We find here not merely a sound and luminous explanation of the chemical principles of the trade, but a notice of numerous matters which have a most important bearing on the successful conduct of alkali works, but which are generally overlooked by even the most experienced technological authors."—*Chemical Review.*

Commercial Chemical Analysis.

THE COMMERCIAL HANDBOOK OF CHEMICAL ANALYSIS; or, Practical Instructions for the determination of the Intrinsic or Commercial Value of Substances used in Manufactures, in Trades, and in the Arts. By A. NORMANDY, Editor of Rose's "Treatise on Chemical Analysis." New Edition, to a great extent Re-written, by HENRY M. NOAD, Ph.D., F.R.S. With numerous Illustrations. Crown 8vo, 12s. 6d. cloth.

"We strongly recommend this book to our readers as a guide, alike indispensable to the housewife as to the pharmaceutical practitioner."—*Medical Times.*

"Essential to the analysts appointed under the new Act. The most recent results are given, and the work is well edited and carefully written."—*Nature.*

Dye-Wares and Colours.

THE MANUAL OF COLOURS AND DYE-WARES: Their Properties, Applications, Valuation, Impurities, and Sophistications. For the use of Dyers, Printers, Drysalters, Brokers, &c. By J. W. SLATER. Second Edition, Revised and greatly Enlarged. Crown 8vo, 7s. 6d. cloth.

"A complete encyclopædia of the *materia tinctoria*. The information given respecting each article is full and precise, and the methods of determining the value of articles such as these, so liable to sophistication, are given with clearness, and are practical as well as valuable."—*Chemist and Druggist.*

"There is no other work which covers precisely the same ground. To students preparing for examinations in dyeing and printing it will prove exceedingly useful."—*Chemical News.*

Pigments.

THE ARTIST'S MANUAL OF PIGMENTS. Showing their Composition, Conditions of Permanency, Non-Permanency, and Adulterations; Effects in Combination with Each Other and with Vehicles; and the most Reliable Tests of Purity. Together with the Science and Arts Department's Examination Questions on Painting. By H. C. STANDAGE. Small crown 8vo, 2s. 6d. cloth.

"This work is indeed *multum-in-parvo*, and we can, with good conscience, recommend it to all who come in contact with pigments, whether as makers, dealers or users."—*Chemical Review.*

"This manual cannot fail to be a very valuable aid to all painters who wish their work to endure and be of a sound character; it is complete and comprehensive."—*Spectator.*

"The author supplies a great deal of very valuable information and memoranda as to the chemical qualities and artistic effect of the principal pigments used by painters."—*Builder.*

AGRICULTURE, LAND MANAGEMENT, etc.

Youatt and Burn's Complete Grazier.
THE COMPLETE GRAZIER, and FARMER'S and CATTLE-BREEDER'S ASSISTANT. A Compendium of Husbandry; especially in the departments connected with the Breeding, Rearing, Feeding, and General Management of Stock; the Management of the Dairy, &c. With Directions for the Culture and Management of Grass Land, of Grain and Root Crops, the Arrangement of Farm Offices, the use of Implements and Machines, and on Draining, Irrigation, Warping, &c.; and the Application and Relative Value of Manures. By WILLIAM YOUATT, Esq., V.S. Twelfth Edition, Enlarged, by ROBERT SCOTT BURN, Author of "Outlines of Modern Farming," "Systematic Small Farming," &c. One large 8vo Volume, 860 pp., with 244 Illustrations, £1 1s. half-bound.

"The standard and text-book with the farmer and grazier."—*Farmers' Magazine.*

"A treatise which will remain a standard work on the subject as long as British agriculture endures."—*Mark Lane Express* (First Notice).

The book deals with all departments of agriculture, and contains an immense amount of valuable information. It is, in fact, an encyclopædia of agriculture put into readable form, and it is the only work equally comprehensive brought down to present date. It deserves a place in the library of every agriculturist."—*Mark Lane Express* (Second Notice)

"This esteemed work is well worthy of a place in the libraries of agriculturists."—*North British Agriculturist.*

Modern Farming.
OUTLINES OF MODERN FARMING. By R. SCOTT BURN. Soils, Manures, and Crops—Farming and Farming Economy—Cattle, Sheep, and Horses—Management of the Dairy, Pigs and Poultry—Utilisation of Town-Sewage, Irrigation, &c. Sixth Edition. In One Vol., 1,250 pp., half-bound, profusely Illustrated, 12s.

"The aim of the author has been to make his work at once comprehensive and trustworthy, and in this aim he has succeeded to a degree which entitles him to much credit."—*Morning Advertiser.*

"Eminently calculated to enlighten the agricultural community on the varied subjects of which it treats, and hence it should find a place in every farmer's library."—*City Press.*

Small Farming.
SYSTEMATIC SMALL FARMING; or, The Lessons of my Farm. Being an Introduction to Modern Farm Practice for Small Farmers in the Culture of Crops; The Feeding of Cattle; The Management of the Dairy, Poultry and Pigs; The Keeping of Farm Work Records; The Ensilage System, Construction of Silos, and other Farm Buildings; The Improvement of Neglected Farms, &c. By ROBERT SCOTT BURN, Author of "Outlines of Landed Estates' Management," and "Outlines of Farm Management," and Editor of "The Complete Grazier." With numerous Illustrations, crown 8vo, 6s. cloth. [*Just published.*

"This is the completest book of its class we have seen, and one which every amateur farmer will read with pleasure and accept as a guide."—*Field.*

"Mr. Scott Burn's pages are severely practical, and the tone of the practical man is felt throughout. The book can only prove a treasure of aid and suggestion to the small farmer of intelligence and energy."—*British Quarterly Review.*

Agricultural Engineering.
FARM ENGINEERING, THE COMPLETE TEXT-BOOK OF. Comprising Draining and Embanking; Irrigation and Water Supply; Farm Roads, Fences, and Gates; Farm Buildings, their arrangement and construction, with plans and estimates; Barn Implements and Machines; Field Implements and Machines; Agricultural Surveying, Levelling, &c. By Prof. JOHN SCOTT, Editor of the *Farmers' Gazette*, late Professor of Agriculture and Rural Economy at the Royal Agricultural College, Cirencester, &c. &c. In One Vol., 1,150 pages, half-bound, with over 600 Illustrations, 12s.

"Written with great care, as well as with knowledge and ability. The author has done his work well; we have found him a very trustworthy guide wherever we have tested his statements. The volume will be of great value to agricultural students, and we have much pleasure in recommending it."—*Mark Lane Express.*

"For a young agriculturist we know of no handy volume so likely to be more usefully studied."—*Bell's Weekly Messenger.*

English Agriculture.

THE FIELDS OF GREAT BRITAIN: A Text-Book of Agriculture, adapted to the Syllabus of the Science and Art Department. For Elementary and Advanced Students. By Hugh Clements (Board of Trade). 18mo, 2s. 6d. cloth.

"A most comprehensive volume, giving a mass of information."—*Agricultural Economist.*

"It is a long time since we have seen a book which has pleased us more, or which contains such a vast and useful fund of knowledge."—*Educational Times.*

Hudson's Land Valuer's Pocket-Book.

THE LAND VALUER'S BEST ASSISTANT: Being Tables on a very much Improved Plan, for Calculating the Value of Estates. With Tables for reducing Scotch, Irish, and Provincial Customary Acres to Statute Measure, &c. By R. Hudson, C.E. New Edition. Royal 32mo, leather, elastic band, 4s.

"This new edition includes tables for ascertaining the value of leases for any term of years; and for showing how to lay out plots of ground of certain acres in forms, square, round, &c., with valuable rules for ascertaining the probable worth of standing timber to any amount; and is of incalculable value to the country gentleman and professional man."—*Farmers' Journal.*

Ewart's Land Improver's Pocket-Book.

THE LAND IMPROVER'S POCKET-BOOK OF FORMULÆ, TABLES and MEMORANDA *required in any Computation relating to the Permanent Improvement of Landed Property.* By John Ewart, Land Surveyor and Agricultural Engineer. Second Edition, Revised. Royal 32mo, oblong, leather, gilt edges, with elastic band, 4s.

"A compendious and handy little volume."—*Spectator.*

Complete Agricultural Surveyor's Pocket-Book.

THE LAND VALUER'S AND LAND IMPROVER'S COMPLETE POCKET-BOOK. Consisting of the above Two Works bound together. Leather, gilt edges, with strap, 7s. 6d.

"Hudson's book is the best ready-reckoner on matters relating to the valuation of land and crops, and its combination with Mr. Ewart's work greatly enhances the value and usefulness of the atter-mentioned. . . . It is most useful as a manual for reference."—*North of England Farmer.*

Farm and Estate Book-keeping.

BOOK-KEEPING FOR FARMERS & ESTATE OWNERS. A Practical Treatise, presenting, in Three Plans, a System adapted to all Classes of Farms. By Johnson M. Woodman, Chartered Accountant. Crown 8vo, 3s. 6d. cloth.

"Will be found of great assistance by those who intend to commence a system of book-keeping, the author's examples being clear and explicit, and his explanations, while full and accurate, being to a large extent free from technicalities."—*Live Stock Journal.*

"The young farmer, land agent and surveyor will find Mr. Woodman's treatise more than repay its cost and study."—*Building News.*

WOODMAN'S YEARLY FARM ACCOUNT BOOK. Giving a Weekly Labour Account and Diary, and showing the Income and Expenditure under each Department of Crops, Live Stock, Dairy, &c. &c. With Valuation, Profit and Loss Account, and Balance Sheet at the end of the Year, and an Appendix of Forms. Folio, 7s. 6d. half-bound.

"Contains every requisite form for keeping farm accounts readily and accurately."—*Agriculture.*

GARDENING, FLORICULTURE, etc.

Early Fruits, Flowers and Vegetables.

THE FORCING GARDEN; or, How to Grow Early Fruits, Flowers, and Vegetables. With Plans, and Estimates for Building Glasshouses, Pits and Frames. Containing also Original Plans for Double Glazing a New Method of Growing the Gooseberry under Glass, &c. &c., and on Ventilation, Protecting Vine Borders, &c. With Illustrations. By Samuel Wood. Crown 8vo, 3s. 6d. cloth.

"A good book, and fairly fills a place that was in some degree vacant."—*Gardeners' Magazine.*

"Mr. Wood's book is an original and exhaustive answer to the question 'How to Grow Early Fruits, Flowers and Vegetables?'"—*Land and Water.*

GARDENING, FLORICULTURE, etc.

Good Gardening.
A PLAIN GUIDE TO GOOD GARDENING ; or, How to Grow Vegetables, Fruits, and Flowers. With Practical Notes on Soils, Manures, Seeds, Planting, Laying-out of Gardens and Grounds, &c. By S. WOOD. Third Edition, with considerable Additions, &c., and numerous Illustrations. Crown 8vo, 5s. cloth.

"A very good book, and one to be highly recommended as a practical guide."—*Athenæum.*

"May be recommended to young gardeners, cottagers, and specially to amateurs, for the plain and trustworthy information it gives on matters too often neglected."—*Gardeners' Chronicle.*

Gainful Gardening.
MULTUM-IN-PARVO GARDENING ; or, How to make One Acre of Land produce £620 a-year by the Cultivation of Fruits and Vegetables ; also, How to Grow Flowers in Three Glass Houses, so as to realise £176 per annum clear Profit. By SAMUEL WOOD, Author of "Good Gardening," &c. Fourth and cheaper Edition, Revised, with Additions. Crown 8vo, 1s. sewed.

"We are bound to recommend it as not only suited to the case of the amateur and gentleman's gardener, but to the market grower."—*Gardeners' Magazine.*

Gardening for Ladies.
THE LADIES' MULTUM-IN-PARVO FLOWER GARDEN, and Amateurs' Complete Guide. With Illustrations. By SAMUEL WOOD. Crown 8vo, 3s. 6d. cloth.

"This volume contains a good deal of sound, common-sense instruction."—*Florist.*

"Full of shrewd hints and useful instructions, based on a lifetime of experience."—*Scotsman.*

Receipts for Gardeners.
GARDEN RECEIPTS. Edited by CHARLES W. QUIN. 12mo, 1s. 6d. cloth limp.

"A useful and handy book, containing a good deal of valuable information."—*Athenæum.*

Kitchen Gardening.
THE KITCHEN AND MARKET GARDEN. By Contributors to "The Garden." Compiled by C. W. SHAW, Editor of "Gardening Illustrated." 12mo, 3s. 6d. cloth boards.

"The most valuable compendium of kitchen and market-garden work published."—*Farmer.*

Cottage Gardening.
COTTAGE GARDENING; or, Flowers, Fruits, and Vegetables for Small Gardens. By E. HOBDAY. 12mo, 1s. 6d. cloth limp.

"Contains much useful information at a small charge."—*Glasgow Herald.*

AUCTIONEERING, ESTATE AGENCY, etc.

Auctioneer's Assistant.
THE APPRAISER, AUCTIONEER, BROKER, HOUSE AND ESTATE AGENT AND VALUER'S POCKET ASSISTANT, for the Valuation for Purchase, Sale, or Renewal of Leases, Annuities and Reversions, and of property generally; with Prices for Inventories, &c. By JOHN WHEELER. Fifth Edition, Extended by C. NORRIS, Valuer, &c. Royal 32mo, 5s. cloth.

"Contains a large quantity of varied and useful information as to the valuation for purchase, sale, or renewal of leases, annuities and reversions, and of property generally, with prices for inventories, and a guide to determine the value of interior fittings and other effects."—*Builder.*

Auctioneering.
AUCTIONEERS: Their Duties and Liabilities. By ROBERT SQUIBBS, Auctioneer. Demy 8vo, 10s. 6d. cloth.

"The position and duties of auctioneers treated compendiously and clearly."—*Builder.*

"Every auctioneer ought to possess a copy of this excellent work."—*Ironmonger*

How to Invest.
HINTS FOR INVESTORS: Being an Explanation of the Mode of Transacting Business on the Stock Exchange. To which are added Comments on the Fluctuations and Table of Quarterly Average prices of Consols since 1759. Also a Copy of the London Daily Stock and Share List. By WALTER M. PLAYFORD, Sworn Broker. Crown 8vo, 2s. cloth.

"An invaluable guide to investors and speculators."—*Bullionist.*

A Complete Epitome of the Laws of this Country.

EVERY MAN'S OWN LAWYER: A Handy-book of the Principles of Law and Equity. By A BARRISTER. Twenty-third Edition. Carefully Revised and brought down to the end of the last Session, including Summaries of the Latest Statute Laws. With Notes and References to the Authorities. Crown 8vo, price 6s. 8d. (saved at every consultation), strongly bound in cloth.

Comprising THE RIGHTS AND WRONGS OF INDIVIDUALS—MERCANTILE AND COMMERCIAL LAW—CRIMINAL LAW—PARISH LAW—COUNTY COURT LAW—GAME AND FISHERY LAWS—POOR MEN'S LAWSUITS—THE LAWS OF BANKRUPTCY—BETS AND WAGERS—CHEQUES, BILLS, AND NOTES—CONTRACTS AND AGREEMENTS—COPYRIGHT —ELECTIONS AND REGISTRATION—INSURANCE—LIBEL AND SLANDER—MARRIAGE AND DIVORCE—MERCHANT SHIPPING—MORTGAGES—SETTLEMENTS—STOCK EXCHANGE PRACTICE—TRADE MARKS AND PATENTS—TRESPASS—NUISANCES, &c.—TRANSFER OF LAND, &c.—WARRANTY—WILLS AND AGREEMENTS, &c. &c.

Opinions of the Press.

"No Englishman ought to be without this book. . . . Any person perfectly uninformed on legal matters, who may require sound information on unknown law points, will, by reference to this book, acquire the necessary information, and thus on many occasions save the expense and loss of time of a visit to a lawyer."—*Engineer.*

"It is a complete code of English Law, written in plain language, which all can understand."— *Weekly Times.*

"A useful and concise epitome of the law, compiled with considerable care."—*Law Magazine.*

"What it professes to be—a complete epitome of the laws of this country, thoroughly intelligible to non-professional readers. The book is a handy one to have in readiness when some knotty point requires ready solution."—*Bell's Life.*

Metropolitan Rating Appeals.

REPORTS OF APPEALS HEARD BEFORE THE COURT OF GENERAL ASSESSMENT SESSIONS, from the Year 1871 to 1885. By EDWARD RYDE and ARTHUR LYON RYDE. Fourth Edition, brought down to the Present Date, with an Introduction to the Valuation (Metropolis) Act, 1869, and an Appendix by WALTER C. RYDE, of the Inner Temple, Barrister-at-Law. 8vo, 16s. cloth.

House Property.

HANDBOOK OF HOUSE PROPERTY: A Popular and Practical Guide to the Purchase, Mortgage, Tenancy, and Compulsory Sale of Houses and Land. By E. L. TARBUCK, Architect and Surveyor. Third Edition, 12mo, 3s. 6d. cloth.

"The advice is thoroughly practical."—*Law Journal.*

"This is a well-written and thoughtful work. We commend the work to the careful study of all interested in questions affecting houses and land."—*Land Agents' Record.*

Inwood's Estate Tables.

TABLES FOR THE PURCHASING OF ESTATES, Freehold, Copyhold, or Leasehold; Annuities, Advowsons, &c., and for the Renewing of Leases held under Cathedral Churches, Colleges, or other Corporate Bodies, for Terms of Years certain, and for Lives; also for Valuing Reversionary Estates, Deferred Annuities, Next Presentations, &c.: together with SMART'S Five Tables of Compound Interest, and an Extension of the same to Lower and Intermediate Rates. By W. INWOOD. 22nd Edition, with considerable Additions, and new and valuable Tables of Logarithms for the more Difficult Computations of the Interest of Money, Discount, Annuities, &c., by M. FEDOR THOMAN, of the Société Crédit Mobilier of Paris. 12mo, 8s. cloth.

"Those interested in the purchase and sale of estates, and in the adjustment of compensation cases, as well as in transactions in annuities, life insurances, &c., will find the present edition of eminent service."—*Engineering.*

"'Inwood's Tables' still maintain a most enviable reputation. The new issue has been enriched by large additional contributions by M. Fedor Thoman, whose carefully arranged Tables cannot fail to be of the utmost utility."—*Mining Journal.*

Agricultural and Tenant-Right Valuation.

THE AGRICULTURAL AND TENANT-RIGHT-VALUER'S ASSISTANT. By TOM BRIGHT, Agricultural Surveyor, Author of "The Live Stock of North Devon," &c. Crown 8vo, 3s. 6d. cloth. [*Just published.*]

"Full of tables and examples in connection with the valuation of tenant-right, estates, labour, contents, and weights of timber, and farm produce of all kinds. The book is well calculated to assist the valuer in the discharge of his duty."—*Agricultural Gazette.*

Weale's Rudimentary Series.

LONDON, 1862.
THE PRIZE MEDAL
Was awarded to the Publishers of
"WEALE'S SERIES."

A NEW LIST OF
WEALE'S SERIES
RUDIMENTARY SCIENTIFIC, EDUCATIONAL, AND CLASSICAL.

Comprising nearly Three Hundred and Fifty *distinct works in almost every department of Science, Art, and Education, recommended to the notice of Engineers, Architects, Builders, Artisans, and Students generally, as well as to those interested in Workmen's Libraries, Literary and Scientific Institutions, Colleges, Schools, Science Classes, &c., &c.*

☞ "WEALE'S SERIES includes Text-Books on almost every branch of Science and Industry, comprising such subjects as Agriculture, Architecture and Building, Civil Engineering, Fine Arts, Mechanics and Mechanical Engineering, Physical and Chemical Science, and many miscellaneous Treatises. The whole are constantly undergoing revision, and new editions, brought up to the latest discoveries in scientific research, are constantly issued. The prices at which they are sold are as low as their excellence is assured."—*American Literary Gazette.*

"Amongst the literature of technical education, WEALE'S SERIES has ever enjoyed a high reputation, and the additions being made by Messrs. CROSBY LOCKWOOD & SON render the series even more complete, and bring the information upon the several subjects down to the present time."—*Mining Journal.*

"It is not too much to say that no books have ever proved more popular with, or more useful to, young engineers and others than the excellent treatises comprised in WEALE'S SERIES."—*Engineer.*

"The excellence of WEALE'S SERIES is now so well appreciated, that it would be wasting our space to enlarge upon their general usefulness and value."—*Builder.*

"WEALE'S SERIES has become a standard as well as an unrivalled collection of treatises in all branches of art and science."—*Public Opinion.*

PHILADELPHIA, 1876.
THE PRIZE MEDAL
Was awarded to the Publishers for
Books: Rudimentary, Scientific,
"WEALE'S SERIES," ETC.

CROSBY LOCKWOOD & SON,

WEALE'S RUDIMENTARY SCIENTIFIC SERIES.

*** The volumes of this Series are freely Illustrated with Woodcuts, or otherwise, where requisite. Throughout the following List it must be understood that the books are bound in limp cloth, unless otherwise stated; *but the volumes marked with a ‡ may also be had strongly bound in cloth boards for 6d. extra.*

N.B.—In ordering from this List it is recommended, as a means of facilitating business and obviating error, to quote the numbers affixed to the volumes, as well as the titles and prices.

CIVIL ENGINEERING, SURVEYING, ETC.

No.
31. *WELLS AND WELL-SINKING.* By JOHN GEO. SWINDELL, A.R.I.B.A., and G. R. BURNELL, C.E. Revised Edition. With a New Appendix on the Qualities of Water. Illustrated. 2s.
35. *THE BLASTING AND QUARRYING OF STONE,* for Building and other Purposes. With Remarks on the Blowing up of Bridges. By Gen. Sir JOHN BURGOYNE, Bart., K.C.B. Illustrated. 1s. 6d.
43. *TUBULAR, AND OTHER IRON GIRDER BRIDGES,* particularly describing the Britannia and Conway Tubular Bridges. By G. DRYSDALE DEMPSEY, C.E. Fourth Edition. 2s.
44. *FOUNDATIONS AND CONCRETE WORKS,* with Practical Remarks on Footings, Sand, Concrete, Béton, Pile-driving, Caissons, and Cofferdams, &c. By E. DOBSON. Fifth Edition. 1s. 6d.
60. *LAND AND ENGINEERING SURVEYING.* By T. BAKER, C.E. New Edition, revised by EDWARD NUGENT, C.E. 2s.‡
80*. *EMBANKING LANDS FROM THE SEA.* With examples and Particulars of actual Embankments, &c. By J. WIGGINS, F.G.S. 2s.
81. *WATER WORKS,* for the Supply of Cities and Towns. With a Description of the Principal Geological Formations of England as influencing Supplies of Water; and Details of Engines and Pumping Machinery for raising Water. By SAMUEL HUGHES, F.G.S., C.E. New Edition. 4s.‡
118. *CIVIL ENGINEERING IN NORTH AMERICA,* a Sketch of. By DAVID STEVENSON, F.R.S.E., &c. Plates and Diagrams. 3s.
167. *IRON BRIDGES, GIRDERS, ROOFS, AND OTHER WORKS.* By FRANCIS CAMPIN, C.E. 2s. 6d.‡
197. *ROADS AND STREETS (THE CONSTRUCTION OF).* By HENRY LAW, C.E., revised and enlarged by D. K. CLARK, C.E., including pavements of Stone, Wood, Asphalte, &c. 4s. 6d.‡
203. *SANITARY WORK IN THE SMALLER TOWNS AND IN VILLAGES.* By C. SLAGG, A.M.I.C.E. Revised Edition. 3s.‡
212. *GAS-WORKS, THEIR CONSTRUCTION AND ARRANGEMENT;* and the Manufacture and Distribution of Coal Gas. Originally written by SAMUEL HUGHES, C.E. Re-written and enlarged by WILLIAM RICHARDS, C.E. Seventh Edition, with important additions. 5s. 6d.‡
213. *PIONEER ENGINEERING.* A Treatise on the Engineering Operations connected with the Settlement of Waste Lands in New Countries. By EDWARD DOBSON, Assoc. Inst. C.E. 4s. 6d.‡
216. *MATERIALS AND CONSTRUCTION;* A Theoretical and Practical Treatise on the Strains, Designing, and Erection of Works of Construction. By FRANCIS CAMPIN, C.E. Second Edition, revised. 3s.‡
219. *CIVIL ENGINEERING.* By HENRY LAW, M.Inst. C.E. Including HYDRAULIC ENGINEERING by GEO. R. BURNELL, M.Inst. C.E. Seventh Edition, revised, with large additions by D. KINNEAR CLARK, M.Inst. C.E. 6s. 6d., Cloth boards, 7s. 6d.

☞ *The ‡ indicates that these vols. may be had strongly bound at 6d. extra.*

LONDON : CROSBY LOCKWOOD AND SON,

MECHANICAL ENGINEERING, ETC.

33. *CRANES*, the Construction of, and other Machinery for Raising Heavy Bodies. By JOSEPH GLYNN, F.R.S. Illustrated. 1s. 6d.
34. *THE STEAM ENGINE.* By Dr. LARDNER. Illustrated. 1s. 6d.
59. *STEAM BOILERS:* their Construction and Management. By R. ARMSTRONG, C.E. Illustrated. 1s. 6d.
82. *THE POWER OF WATER*, as applied to drive Flour Mills, and to give motion to Turbines, &c. By JOSEPH GLYNN, F.R.S. 2s.‡
98. *PRACTICAL MECHANISM*, the Elements of; and Machine Tools. By T. BAKER, C.E. With Additions by J. NASMYTH, C.E. 2s. 6d.‡
139. *THE STEAM ENGINE*, a Treatise on the Mathematical Theory of, with Rules and Examples for Practical Men. By T. BAKER, C.E. 1s. 6d.
164. *MODERN WORKSHOP PRACTICE*, as applied to Steam Engines, Bridges, Cranes, Ship-building, &c. By J. G. WINTON. 3s.‡
165. *IRON AND HEAT*, exhibiting the Principles concerned in the Construction of Iron Beams, Pillars, and Girders. By J. ARMOUR. 2s. 6d.‡
166. *POWER IN MOTION:* Horse-Power, Toothed-Wheel Gearing, Long and Short Driving Bands, and Angular Forces. By J. ARMOUR, 2s. 6d.‡
171. *THE WORKMAN'S MANUAL OF ENGINEERING DRAWING.* By J. MAXTON. 6th Edn. With 7 Plates and 350 Cuts. 3s. 6d.‡
190. *STEAM AND THE STEAM ENGINE*, Stationary and Portable. By JOHN SEWELL and D. K. CLARK, M.I.C.E. 3s. 6d.‡
200. *FUEL*, its Combustion and Economy. By C. W. WILLIAMS, With Recent Practice in the Combustion and Economy of Fuel—Coal, Coke, Wood, Peat, Petroleum, &c.—by D. K. CLARK, M.I.C.E. 3s. 6d.‡
202. *LOCOMOTIVE ENGINES.* By G. D. DEMPSEY, C.E.; with large additions by D. KINNEAR CLARK, M.I.C.E. 3s.‡
211. *THE BOILERMAKER'S ASSISTANT* in Drawing, Templating, and Calculating Boiler and Tank Work. By JOHN COURTNEY, Practical Boiler Maker. Edited by D. K. CLARK, C.E. 100 Illustrations. 2s.
217. *SEWING MACHINERY:* Its Construction, History, &c., with full Technical Directions for Adjusting, &c. By J. W. URQUHART, C.E. 2s.‡
223. *MECHANICAL ENGINEERING.* Comprising Metallurgy, Moulding, Casting, Forging, Tools, Workshop Machinery, Manufacture of the Steam Engine, &c. By FRANCIS CAMPIN, C.E. 2s. 6d.‡
236. *DETAILS OF MACHINERY.* Comprising Instructions for the Execution of various Works in Iron. By FRANCIS CAMPIN, C.E. 3s.‡
237. *THE SMITHY AND FORGE;* including the Farrier's Art and Coach Smithing. By W. J. E. CRANE. Illustrated. 2s. 6d.‡
238. *THE SHEET-METAL WORKER'S GUIDE;* a Practical Handbook for Tinsmiths, Coppersmiths, Zincworkers, &c. With 94 Diagrams and Working Patterns. By W. J. E. CRANE. 1s. 6d.
251. *STEAM AND MACHINERY MANAGEMENT:* with Hints on Construction and Selection. By M. POWIS BALE, M.I.M.E. 2s. 6d.‡
254. *THE BOILERMAKER'S READY-RECKONER.* By J. COURTNEY. Edited by D. K. CLARK, C.E. 4s.; limp; 5s., half-bound.
255. *LOCOMOTIVE ENGINE-DRIVING.* A Practical Manual for Engineers in charge of Locomotive Engines. By MICHAEL REYNOLDS, M.S.E. Seventh Edition. 3s. 6d., limp; 4s. 6d. cloth boards.
256. *STATIONARY ENGINE-DRIVING.* A Practical Manual for Engineers in charge of Stationary Engines. By MICHAEL REYNOLDS, M.S.E. Third Edition. 3s. 6d. limp; 4s. 6d. cloth boards.
260. *IRON BRIDGES OF MODERATE SPAN:* their Construction and Erection. By HAMILTON W. PENDRED, late Inspector of Ironwork to the Salford Corporation. 2s. [*Just published.*

☞ The ‡ indicates that these vols. may be had strongly bound at 6d. extra.

7, STATIONERS' HALL COURT, LUDGATE HILL, E.C.

MINING, METALLURGY, ETC.

4. *MINERALOGY*, Rudiments of; a concise View of the General Properties of Minerals. By A. RAMSAY, F.G.S., F.R.G.S., &c. Third Edition, revised and enlarged. Illustrated. 3s. 6d.‡

117. *SUBTERRANEOUS SURVEYING*, with and without the Magnetic Needle. By T. FENWICK and T. BAKER, C.E. Illustrated. 2s. 6d. ‡

133. *METALLURGY OF COPPER* ; an Introduction to the Methods of Seeking, Mining, and Assaying Copper, and Manufacturing its Alloys. By ROBERT H. LAMBORN. Ph.D. Woodcuts. 2s. 6d.‡

135. *ELECTRO-METALLURGY;* Practically Treated. By ALEXANDER WATT, F.R.S.S.A. Eighth Edition, revised, with additional Matter and Illustrations, including the most recent Processes. 3s.‡

172. *MINING TOOLS*, Manual of. For the Use of Mine Managers, Agents, Students, &c. By WILLIAM MORGANS. 2s. 6d.‡

172*. *MINING TOOLS, ATLAS* of Engravings to Illustrate the above, containing 235 Illustrations, drawn to Scale. 4to. 4s. 6d.; cloth boards, 6s.

176. *METALLURGY OF IRON.* Containing History of Iron Manufacture, Methods of Assay, and Analyses of Iron Ores, Processes of Manufacture of Iron and Steel, &c. By H. BAUERMAN, F.G.S. Fifth Edition, revised and enlarged. 5s.‡

180. *COAL AND COAL MINING.* By WARINGTON W. SMYTH, M.A., F.R.S. Sixth Edition, revised. 3s. 6d.‡

195. *THE MINERAL SURVEYOR AND VALUER'S COMPLETE GUIDE*, with new Traverse Tables, and Descriptions of Improved Instruments ; also the Correct Principles of Laying out and Valuing Mineral Properties. By WILLIAM LINTERN, Mining and Civil Engineer. 3s. 6d.‡

214. *SLATE AND SLATE QUARRYING*, Scientific, Practical, and Commercial. By D. C. DAVIES, F.G.S., Mining Engineer, &c. 3s.‡

220. *MAGNETIC SURVEYING, AND ANGULAR SURVEYING*, with Records of the Peculiarities of Needle Disturbances. Compiled from the Results of carefully made Experiments. By W. LINTERN. 2s.

ARCHITECTURE, BUILDING, ETC.

16. *ARCHITECTURE—ORDERS*—The Orders and their Æsthetic Principles. By W. H. LEEDS. Illustrated. 1s. 6d.

17. *ARCHITECTURE—STYLES*—The History and Description of the Styles of Architecture of Various Countries, from the Earliest to the Present Period. By T. TALBOT BURY, F.R.I.B.A., &c. Illustrated. 2s.
*** ORDERS AND STYLES OF ARCHITECTURE, *in One Vol.*, 3s. 6d.

18. *ARCHITECTURE—DESIGN*—The Principles of Design in Architecture, as deducible from Nature and exemplified in the Works of the Greek and Gothic Architects. By E. L. GARBETT, Architect. Illustrated. 2s.6d.
*** *The three preceding Works, in One handsome Vol., half bound, entitled* "MODERN ARCHITECTURE," *price 6s.*

22. *THE ART OF BUILDING*, Rudiments of. General Principles of Construction, Materials used in Building, Strength and Use of Materials, Working Drawings, Specifications, and Estimates. By E. DOBSON, 2s.‡

25. *MASONRY AND STONECUTTING :* Rudimentary Treatise on the Principles of Masonic Projection and their application to Construction. By EDWARD DOBSON, M.R.I.B.A., &c. 2s. 6d.‡

42. *COTTAGE BUILDING.* By C. BRUCE ALLEN, Architect. Tenth Edition, revised and enlarged. With a Chapter on Economic Cottages for Allotments, by EDWARD E. ALLEN, C.E. 2s.

45. *LIMES, CEMENTS, MORTARS, CONCRETES, MASTICS,* PLASTERING &c. By G. R. BURNELL, C.E. Twelfth Edition. 1s. 6d.

☞ *The* ‡ *indicates that these vols. may be had strongly bound at 6d. extra.*

LONDON : CROSBY LOCKWOOD AND SON,

Architecture, Building, etc., *continued.*

57. WARMING AND VENTILATION. An Exposition of the General Principles as applied to Domestic and Public Buildings, Mines, Lighthouses, Ships, &c. By C. TOMLINSON, F.R.S., &c. Illustrated. 3s.

111. ARCHES, PIERS, BUTTRESSES, &c.: Experimental Essays on the Principles of Construction. By W. BLAND. Illustrated. 1s. 6d.

116. THE ACOUSTICS OF PUBLIC BUILDINGS; or, The Principles of the Science of Sound applied to the purposes of the Architect and Builder. By T. ROGER SMITH, M.R.I.B.A., Architect. Illustrated. 1s. 6d.

127. ARCHITECTURAL MODELLING IN PAPER, the Art of. By T. A. RICHARDSON, Architect. Illustrated. 1s. 6d.

128. VITRUVIUS—THE ARCHITECTURE OF MARCUS VITRUVIUS POLLO. In Ten Books. Translated from the Latin by JOSEPH GWILT, F.S.A., F.R.A.S. With 23 Plates. 5s.

130. GRECIAN ARCHITECTURE, An Inquiry into the Principles of Beauty in; with an Historical View of the Rise and Progress of the Art in Greece. By the EARL OF ABERDEEN. 1s.

** *The two preceding Works in One handsome Vol., half bound, entitled* "ANCIENT ARCHITECTURE," *price 6s.*

132. THE ERECTION OF DWELLING-HOUSES. Illustrated by a Perspective View, Plans, Elevations, and Sections of a pair of Semi-detached Villas, with the Specification, Quantities, and Estimates, &c. By S. H. BROOKS. New Edition, with Plates. 2s. 6d.‡

156. QUANTITIES & MEASUREMENTS in Bricklayers', Masons', Plasterers', Plumbers', Painters', Paperhangers', Gilders', Smiths', Carpenters' and Joiners' Work. By A. C. BEATON, Surveyor. New Edition. 1s. 6d.

175. LOCKWOOD & CO.'S BUILDER'S AND CONTRACTOR'S PRICE BOOK, containing the latest Prices of all kinds of Builders' Materials and Labour, and of all Trades connected with Building, &c., &c. Edited by F. T. W. MILLER, Architect. Published annually. 3s. 6d.; half bound, 4s.

182. CARPENTRY AND JOINERY—THE ELEMENTARY PRINCIPLES OF CARPENTRY. Chiefly composed from the Standard Work of THOMAS TREDGOLD, C.E. With a TREATISE ON JOINERY by E. WYNDHAM TARN, M.A. Fourth Edition, Revised. 3s. 6d.‡

182*. CARPENTRY AND JOINERY. ATLAS of 35 Plates to accompany the above. With Descriptive Letterpress. 4to. 6s.

185. THE COMPLETE MEASURER; the Measurement of Boards, Glass, &c.; Unequal-sided, Square-sided, Octagonal-sided, Round Timber and Stone, and Standing Timber, &c. By RICHARD HORTON. Fifth Edition. 4s.; strongly bound in leather, 5s.

187. HINTS TO YOUNG ARCHITECTS. By G. WIGHTWICK. New Edition. By G. H. GUILLAUME. Illustrated. 3s. 6d.‡

188. HOUSE PAINTING, GRAINING, MARBLING, AND SIGN WRITING: with a Course of Elementary Drawing for House-Painters, Sign-Writers, &c., and a Collection of Useful Receipts. By ELLIS A. DAVIDSON. Fourth Edition. With Coloured Plates. 5s. cloth limp; 6s. cloth boards.

189. THE RUDIMENTS OF PRACTICAL BRICKLAYING. In Six Sections: General Principles; Arch Drawing, Cutting, and Setting; Pointing; Paving, Tiling, Materials; Slating and Plastering; Practical Geometry, Mensuration, &c. By ADAM HAMMOND. Sixth Edition. 1s. 6d.

191. PLUMBING. A Text-Book to the Practice of the Art or Craft of the Plumber. With Chapters upon House Drainage. Fourth Edition. With 330 Illustrations. By W. P. BUCHAN. 3s. 6d.‡

192. THE TIMBER IMPORTER'S, TIMBER MERCHANT'S, and BUILDER'S STANDARD GUIDE. By R. E. GRANDY. 3s.‡

206. A BOOK ON BUILDING, Civil and Ecclesiastical, including CHURCH RESTORATION. With the Theory of Domes and the Great Pyramid, &c. By Sir EDMUND BECKETT, Bart., LL.D., Q.C., F.R.A.S. 4s. 6d.‡

☞ *The ‡ indicates that these vols. may be had strongly bound at 6d. extra.*

7, STATIONERS' HALL COURT, LUDGATE HILL, E.C.

Architecture, Building, etc., *continued*.

226. *THE JOINTS MADE AND USED BY BUILDERS* in the Construction of various kinds of Engineering and Architectural Works. By WYVILL J. CHRISTY, Architect. With upwards of 160 Engravings on Wood. 3s.‡

228. *THE CONSTRUCTION OF ROOFS OF WOOD AND IRON* By E. WYNDHAM TARN, M.A., Architect. Second Edition, revised. 1s. 6d

229. *ELEMENTARY DECORATION:* as applied to the Interior and Exterior Decoration of Dwelling-Houses, &c. By J. W. FACEY. 2s.

257. *PRACTICAL HOUSE DECORATION.* A Guide to the Art of Ornamental Painting, the Arrangement of Colours in Apartments, and the Principles of Decorative Design. With Remarks on the Nature and Properties of Pigments. By JAMES W. FACEY. 2s. 6d.

*** *The two preceding Works, in One handsome Vol., half-bound, entitled* "HOUSE DECORATION, ELEMENTARY AND PRACTICAL," *price* 5s.

230. *HANDRAILING.* Showing New and Simple Methods for finding the Pitch of the Plank. Drawing the Moulds, Bevelling, Jointing-up, and Squaring the Wreath. By GEORGE COLLINGS. Plates and Diagrams. 1s. 6d.

247. *BUILDING ESTATES:* a Rudimentary Treatise on the Development, Sale, Purchase, and General Management of Building Land. By FOWLER MAITLAND, Surveyor. Illustrated. 2s.

248. *PORTLAND CEMENT FOR USERS.* By HENRY FAIJA, Assoc. M. Inst. C.E. Second Edition, corrected. Illustrated. 2s.

252. *BRICKWORK:* a Practical Treatise, embodying the General and Higher Principles of Bricklaying, Cutting and Setting, &c. By F. WALKER. Second Edition, Revised and Enlarged. 1s. 6d.

23, 189, 252. *THE PRACTICAL BRICK AND TILE BOOK.* Comprising: BRICK AND TILE MAKING, by E. DOBSON, A.I.C.E.; PRACTICAL BRICKLAYING, by A. HAMMOND; BRICKWORK, by F. WALKER. 550 pp. with 270 Illustrations. 6s. Strongly half-bound.

253. *THE TIMBER MERCHANT'S, SAW-MILLER'S, AND IMPORTER'S FREIGHT-BOOK AND ASSISTANT.* By WM. RICHARDSON. With a Chapter on Speeds of Saw-Mill Machinery, &c. By M. POWIS BALE, A.M.Inst.C.E. 3s.‡

250. *CIRCULAR WORK IN CARPENTRY AND JOINERY.* A Practical Treatise on Circular Work of Single and Double Curvature. By GEORGE COLLINGS, Author of "A Practical Treatise on Handrailing." 2s. 6d. [*Just published.*

259. *GAS FITTING:* A Practical Handbook treating of every Description of Gas Laying and Fitting. By JOHN BLACK. With 122 Illustrations. 2s. 6d.‡ [*Just published.*

261. *SHORING AND ITS APPLICATION:* A Handbook for the Use of Students. By GEORGE H. BLAGROVE. 1s. 6d. [*Just published.*

SHIPBUILDING, NAVIGATION, MARINE ENGINEERING, ETC.

51. *NAVAL ARCHITECTURE.* An Exposition of the Elementary Principles of the Science, and their Practical Application to Naval Construction. By J. PEAKE. Fifth Edition, with Plates and Diagrams. 3s. 6d.‡

53*. *SHIPS FOR OCEAN & RIVER SERVICE*, Elementary and Practical Principles of the Construction of. By H. A. SOMMERFELDT. 1s. 6d.

53**. *AN ATLAS OF ENGRAVINGS* to Illustrate the above. Twelve large folding plates. Royal 4to, cloth. 7s. 6d.

54. *MASTING, MAST-MAKING, AND RIGGING OF SHIPS*, Also Tables of Spars, Rigging, Blocks; Chain, Wire, and Hemp Ropes, &c., relative to every class of vessels. By ROBERT KIPPING, N.A. 2s.‡

54*. *IRON SHIP-BUILDING.* With Practical Examples and Details. By JOHN GRANTHAM, C.E. 5th Edition. 4s.

☞ *The ‡ indicates that these vols. may be had strongly bound at 6d. extra.*

LONDON: CROSBY LOCKWOOD AND SON,

Shipbuilding, Navigation, Marine Engineering, etc., *cont.*

55. *THE SAILOR'S SEA BOOK:* a Rudimentary Treatise on Navigation. By JAMES GREENWOOD, B.A. With numerous Woodcuts and Coloured Plates. New and enlarged edition. By W. H. ROSSER. 2s. 6d.‡

80. *MARINE ENGINES AND STEAM VESSELS.* By ROBERT MURRAY, C.E. Eighth Edition, thoroughly Revised, with Additions by the Author and by GEORGE CARLISLE, C.E., Senior Surveyor to the Board of Trade, Liverpool. 4s. 6d. limp ; 5s. cloth boards. [*Just published.*

83*bis.* *THE FORMS OF SHIPS AND BOATS.* By W. BLAND. Seventh Edition, Revised, with numerous Illustrations and Models. 1s. 6d.

99. *NAVIGATION AND NAUTICAL ASTRONOMY,* in Theory and Practice. By Prof. J. R. YOUNG. New Edition. 2s. 6d.

106. *SHIPS' ANCHORS,* a Treatise on. By G. COTSELL, N.A. 1s. 6d.

149. *SAILS AND SAIL-MAKING.* With Draughting, and the Centre of Effort of the Sails; Weights and Sizes of Ropes ; Masting, Rigging, and Sails of Steam Vessels, &c. 11th Edition. By R. KIPPING, N.A., 2s. 6d.‡

155. *ENGINEER'S GUIDE TO THE ROYAL & MERCANTILE NAVIES.* By a PRACTICAL ENGINEER. Revised by D. F. M'CARTHY. 3s.

55 & 204. *PRACTICAL NAVIGATION.* Consisting of The Sailor's Sea-Book. By JAMES GREENWOOD and W. H. ROSSER. Together with the requisite Mathematical and Nautical Tables for the Working of the Problems. By H. LAW, C.E., and Prof. J. R. YOUNG. 7s. Half-bound.

AGRICULTURE, GARDENING, ETC.

61*. *A COMPLETE READY RECKONER FOR THE ADMEA-SUREMENT OF LAND,* &c. By A. ARMAN. Second Edition, revised and extended by C. NORRIS, Surveyor, Valuer, &c. 2s.

131. *MILLER'S, CORN MERCHANT'S, AND FARMER'S READY RECKONER.* Second Edition, with a Price List of Modern Flour-Mill Machinery, by W. S. HUTTON, C.E. 2s.

140. *SOILS, MANURES, AND CROPS.* (Vol. 1. OUTLINES OF MODERN FARMING.) By R. SCOTT BURN. Woodcuts. 2s.

141. *FARMING & FARMING ECONOMY,* Notes, Historical and Practical, on. (Vol. 2. OUTLINES OF MODERN FARMING.) By R. SCOTT BURN. 3s.

142. *STOCK; CATTLE, SHEEP, AND HORSES.* (Vol. 3. OUTLINES OF MODERN FARMING.) By R. SCOTT BURN. Woodcuts. 2s. 6d.

145. *DAIRY, PIGS, AND POULTRY,* Management of the. By R. SCOTT BURN. (Vol. 4. OUTLINES OF MODERN FARMING.) 2s.

146. *UTILIZATION OF SEWAGE, IRRIGATION, AND RECLAMATION OF WASTE LAND.* (Vol. 5. OUTLINES OF MODERN FARMING.) By R. SCOTT BURN. Woodcuts. 2s. 6d.

*** *Nos.* 140-1-2-5-6, *in One Vol., handsomely half-bound, entitled* "OUTLINES OF MODERN FARMING." *By* ROBERT SCOTT BURN. *Price* 12s.

177. *FRUIT TREES,* The Scientific and Profitable Culture of. From the French of DU BREUIL. Revised by GEO. GLENNY. 187 Woodcuts. 3s. 6d.‡

198. *SHEEP; THE HISTORY, STRUCTURE, ECONOMY, AND DISEASES OF.* By W. C. SPOONER, M.R.V.C., &c. Fourth Edition, enlarged, including Specimens of New and Improved Breeds. 3s. 6d.‡

201. *KITCHEN GARDENING MADE EASY.* By GEORGE M. F. GLENNY. Illustrated. 1s. 6d.‡

207. *OUTLINES OF FARM MANAGEMENT, and the Organization of Farm Labour.* By R. SCOTT BURN. 2s. 6d.‡

208. *OUTLINES OF LANDED ESTATES MANAGEMENT.* By R. SCOTT BURN. 2s. 6d.‡

*** *Nos.* 207 *&* 208 *in One Vol., handsomely half-bound, entitled* "OUTLINES OF LANDED ESTATES AND FARM MANAGEMENT." *By* R. SCOTT BURN. *Price* 6s.

☞ *The* ‡ *indicates that these vols. may be had strongly bound at* 6d. *extra.*

7, STATIONERS' HALL COURT, LUDGATE HILL, E.C.

Agriculture, Gardening, etc., *continued.*

209. **THE TREE PLANTER AND PLANT PROPAGATOR.** A Practical Manual on the Propagation of Forest Trees, Fruit Trees, Flowering Shrubs, Flowering Plants, &c. By SAMUEL WOOD. 2s.‡

210. **THE TREE PRUNER.** A Practical Manual on the Pruning of Fruit Trees, including also their Training and Renovation; also the Pruning of Shrubs, Climbers, and Flowering Plants. By SAMUEL WOOD. 2s.‡

*** *Nos.* 209 & 210 *in One Vol., handsomely half-bound, entitled* "THE TREE PLANTER, PROPAGATOR, AND PRUNER." By SAMUEL WOOD. *Price* 5s.

218. **THE HAY AND STRAW MEASURER:** Being New Tables for the Use of Auctioneers, Valuers, Farmers, Hay and Straw Dealers, &c. By JOHN STEELE. Fourth Edition. 2s.

222. **SUBURBAN FARMING.** The Laying-out and Cultivation of Farms, adapted to the Produce of Milk, Butter, and Cheese, Eggs, Poultry, and Pigs. By Prof. JOHN DONALDSON and R. SCOTT BURN. 3s. 6d.‡

231. **THE ART OF GRAFTING AND BUDDING.** By CHARLES BALTET. With Illustrations. 2s. 6d.‡

232. **COTTAGE GARDENING;** or, Flowers, Fruits, and Vegetables for Small Gardens. By E. HOBDAY. 1s. 6d.

233. **GARDEN RECEIPTS.** Edited by CHARLES W. QUIN. 1s. 6d.

234. **THE KITCHEN AND MARKET GARDEN.** Compiled by C. W. SHAW, Editor of "Gardening Illustrated." 3s.‡

239. **DRAINING AND EMBANKING.** A Practical Treatise, embodying the most recent experience in the Application of Improved Methods. By JOHN SCOTT, late Professor of Agriculture and Rural Economy at the Royal Agricultural College, Cirencester. With 68 Illustrations. 1s. 6d.

240. **IRRIGATION AND WATER SUPPLY.** A Treatise on Water Meadows, Sewage Irrigation, Warping, &c.; on the Construction of Wells, Ponds, and Reservoirs; and on Raising Water by Machinery for Agricultural and Domestic Purposes. By Prof. JOHN SCOTT. With 34 Illus. 1s. 6d.

241. **FARM ROADS, FENCES, AND GATES.** A Practical Treatise on the Roads, Tramways, and Waterways of the Farm; the Principles of Enclosures; and the different kinds of Fences, Gates, and Stiles. By Professor JOHN SCOTT. With 75 Illustrations. 1s. 6d.

242. **FARM BUILDINGS.** A Practical Treatise on the Buildings necessary for various kinds of Farms, their Arrangement and Construction, including Plans and Estimates. By Prof. JOHN SCOTT. With 105 Illus. 2s.

243. **BARN IMPLEMENTS AND MACHINES.** A Practical Treatise on the Application of Power to the Operations of Agriculture; and on various Machines used in the Threshing-barn, in the Stock-yard, and in the Dairy, &c. By Prof. J. SCOTT. With 123 Illustrations. 2s.

244. **FIELD IMPLEMENTS AND MACHINES.** A Practical Treatise on the Varieties now in use, with Principles and Details of Construction, their Points of Excellence, and Management. By Professor JOHN SCOTT. With 138 Illustrations. 2s.

245. **AGRICULTURAL SURVEYING.** A Practical Treatise on Land Surveying, Levelling, and Setting-out; and on Measuring and Estimating Quantities, Weights, and Values of Materials, Produce, Stock, &c. By Prof. JOHN SCOTT. With 62 Illustrations. 1s. 6d.

*** *Nos.* 239 *to* 245 *in One Vol., handsomely half-bound, entitled* "THE COMPLETE TEXT-BOOK OF FARM ENGINEERING." By Professor JOHN SCOTT. *Price* 12s.

250. **MEAT PRODUCTION.** A Manual for Producers, Distributors, &c. By JOHN EWART. 2s. 6d.‡

☞ *The* ‡ *indicates that these vols. may be had strongly bound at 6d. extra.*

LONDON : CROSBY LOCKWOOD AND SON,

MATHEMATICS, ARITHMETIC, ETC.

32. *MATHEMATICAL INSTRUMENTS*, a Treatise on; in which their Construction and the Methods of Testing, Adjusting, and Using them are concisely Explained. By J. F. HEATHER, M.A., of the Royal Military Academy, Woolwich. Original Edition, in 1 vol., Illustrated. 1s. 6d.

** In ordering the above, be careful to say, "*Original Edition*" (*No.* 32), *to distinguish it from the Enlarged Edition in* 3 *vols.* (*Nos.* 168-9-70.)

76. *DESCRIPTIVE GEOMETRY*, an Elementary Treatise on; with a Theory of Shadows and of Perspective, extracted from the French of G. MONGE. To which is added, a description of the Principles and Practice of Isometrical Projection. By J. F. HEATHER, M.A. With 14 Plates. 2s.

178. *PRACTICAL PLANE GEOMETRY:* giving the Simplest Modes of Constructing Figures contained in one Plane and Geometrical Construction of the Ground. By . F. HEATHER, M.A. With 215 Woodcuts. 2s.

83. *COMMERCIAL BOOK-KEEPING*. With Commercial Phrases and Forms in English, French, Italian, and German. By JAMES HADDON, M.A., Arithmetical Master of King's College School, London. 1s. 6d.

84. *ARITHMETIC*, a Rudimentary Treatise on: with full Explanations of its Theoretical Principles, and numerous Examples for Practice. By Professor J. R. YOUNG. Tenth Edition, corrected. 1s. 6d.

84*. A KEY to the above, containing Solutions in full to the Exercises, together with Comments, Explanations, and Improved Processes, for the Use of Teachers and Unassisted Learners. By J. R. YOUNG. 1s. 6d.

85. *EQUATIONAL ARITHMETIC*, applied to Questions of Interest, Annuities, Life Assurance, and General Commerce; with various Tables by which all Calculations may be greatly facilitated. By W. HIPSLEY. 2s.

86. *ALGEBRA*, the Elements of. By JAMES HADDON, M.A. With Appendix, containing miscellaneous Investigations, and a Collection of Problems in various parts of Algebra. 2s.

86*. A KEY AND COMPANION to the above Book, forming an extensive repository of Solved Examples and Problems in Illustration of the various Expedients necessary in Algebraical Operations. By J. R. YOUNG. 1s. 6d.

88. *EUCLID*, THE ELEMENTS OF: with many additional Propositions
89. and Explanatory Notes: to which is prefixed, an Introductory Essay on Logic. By HENRY LAW, C.E. 2s. 6d.‡

** Sold also separately, viz.:—

88. EUCLID, The First Three Books. By HENRY LAW, C.E. 1s. 6d.
89. EUCLID, Books 4, 5, 6, 11, 12. By HENRY LAW, C.E. 1s. 6d.

90. *ANALYTICAL GEOMETRY AND CONIC SECTIONS*, By JAMES HANN. A New Edition, by Professor J. R. YOUNG. 2s.‡

91. *PLANE TRIGONOMETRY*, the Elements of. By JAMES HANN, formerly Mathematical Master of King's College, London. 1s. 6d.

92. *SPHERICAL TRIGONOMETRY*, the Elements of. By JAMES HANN. Revised by CHARLES H. DOWLING, C.E. 1s.

** Or with "The Elements of Plane Trigonometry," in One Volume, 2s. 6d.

93. *MENSURATION AND MEASURING*. With the Mensuration and Levelling of Land for the Purposes of Modern Engineering. By T. BAKER, C.E. New Edition by E. NUGENT, C.E. Illustrated. 1s. 6d.

101. *DIFFERENTIAL CALCULUS*, Elements of the. By W. S. B. WOOLHOUSE, F.R.A.S., &c. 1s. 6d.

102. *INTEGRAL CALCULUS*, Rudimentary Treatise on the. By HOMERSHAM COX, B.A. Illustrated. 1s.

105. *MNEMONICAL LESSONS.* — GEOMETRY, ALGEBRA, AND TRIGONOMETRY, in Easy Mnemonical Lessons. By the Rev. THOMAS PENYNGTON KIRKMAN, M.A. 1s. 6d.

136. *ARITHMETIC*, Rudimentary, for the Use of Schools and Self-Instruction. By JAMES HADDON, M.A. Revised by A. ARMAN. 1s. 6d.

137. A KEY TO HADDON'S RUDIMENTARY ARITHMETIC. By A. ARMAN. 1s. 6d.

☞ *The* ‡ *indicates that these vols. may be had strongly bound at* 6d. *extra.*

7, STATIONERS' HALL COURT, LUDGATE HILL, E.C.

Mathematics, Arithmetic, etc., *continued.*

168. **DRAWING AND MEASURING INSTRUMENTS.** Including—I. Instruments employed in Geometrical and Mechanical Drawing, and in the Construction, Copying, and Measurement of Maps and Plans. II. Instruments used for the purposes of Accurate Measurement, and for Arithmetical Computations. By J. F. HEATHER, M.A. Illustrated. 1s. 6d.

169. **OPTICAL INSTRUMENTS.** Including (more especially) Telescopes, Microscopes, and Apparatus for producing copies of Maps and Plans by Photography. By J. F. HEATHER, M.A. Illustrated. 1s. 6d.

170. **SURVEYING AND ASTRONOMICAL INSTRUMENTS.** Including—I. Instruments Used for Determining the Geometrical Features of a portion of Ground. II. Instruments Employed in Astronomical Observations. By J. F. HEATHER, M.A. Illustrated. 1s. 6d.

⁂ *The above three volumes form an enlargement of the Author's original work,* "*Mathematical Instruments.*" (*See No. 32 in the Series.*)

168.⎫
169.⎬ **MATHEMATICAL INSTRUMENTS.** By J. F. HEATHER, M.A. Enlarged Edition, for the most part entirely re-written. The 3 Parts as
170.⎭ above, in One thick Volume. With numerous Illustrations. 4s. 6d.‡

158. **THE SLIDE RULE, AND HOW TO USE IT;** containing full, easy, and simple Instructions to perform all Business Calculations with unexampled rapidity and accuracy. By CHARLES HOARE, C.E. Fifth Edition. With a Slide Rule in tuck of cover. 2s. 6d.‡

196. **THEORY OF COMPOUND INTEREST AND ANNUITIES;** with Tables of Logarithms for the more Difficult Computations of Interest, Discount, Annuities, &c. By FÉDOR THOMAN. 4s.‡

199. **THE COMPENDIOUS CALCULATOR;** or, Easy and Concise Methods of Performing the various Arithmetical Operations required in Commercial and Business Transactions; together with Useful Tables. By D. O'GORMAN. Twenty-sixth Edition, carefully revised by C. NORRIS. 3s., cloth limp; 3s. 6d., strongly half-bound in leather.

204. **MATHEMATICAL TABLES,** for Trigonometrical, Astronomical, and Nautical Calculations; to which is prefixed a Treatise on Logarithms. By HENRY LAW, C.E. Together with a Series of Tables for Navigation and Nautical Astronomy. By Prof. J. R. YOUNG. New Edition. 4s.‡

204*. **LOGARITHMS.** With Mathematical Tables for Trigonometrical, Astronomical, and Nautical Calculations. By HENRY LAW, M.Inst.C.E. New and Revised Edition. (Forming part of the above Work). 3s.

221. **MEASURES, WEIGHTS, AND MONEYS OF ALL NATIONS,** and an Analysis of the Christian, Hebrew, and Mahometan Calendars. By W. S. B. WOOLHOUSE, F.R.A.S., F.S.S. Sixth Edition. 2s.‡

227. **MATHEMATICS AS APPLIED TO THE CONSTRUCTIVE ARTS.** Illustrating the various processes of Mathematical Investigation, by means of Arithmetical and Simple Algebraical Equations and Practical Examples. By FRANCIS CAMPIN, C.E. Second Edition. 3s.‡

PHYSICAL SCIENCE, NATURAL PHILOSOPHY, ETC.

1. **CHEMISTRY.** By Professor GEORGE FOWNES, F.R.S. With an Appendix on the Application of Chemistry to Agriculture. 1s.

2. **NATURAL PHILOSOPHY,** Introduction to the Study of. By C. TOMLINSON. Woodcuts. 1s. 6d.

6. **MECHANICS,** Rudimentary Treatise on. By CHARLES TOMLINSON. Illustrated. 1s. 6d.

7. **ELECTRICITY;** showing the General Principles of Electrical Science, and the purposes to which it has been applied. By Sir W. SNOW HARRIS, F.R.S., &c. With Additions by R. SABINE, C.E., F.S.A. 1s. 6d.

7*. **GALVANISM.** By Sir W. SNOW HARRIS. New Edition by ROBERT SABINE, C.E., F.S.A. 1s. 6d.

8. **MAGNETISM;** being a concise Exposition of the General Principles of Magnetical Science. By Sir W. SNOW HARRIS. New Edition, revised by H. M. NOAD, Ph.D. With 165 Woodcuts. 3s. 6d.‡

☞ *The ‡ indicates that these vols. may be had strongly bound at 6d. extra.*

LONDON : CROSBY LOCKWOOD AND SON,

Physical Science, Natural Philosophy, etc., *continued*.

11. *THE ELECTRIC TELEGRAPH;* its History and Progress; with Descriptions of some of the Apparatus. By R. SABINE, C.E., F.S.A. 3s.
12. *PNEUMATICS,* for the Use of Beginners. By CHARLES TOMLINSON. Illustrated. 1s. 6d.
72. *MANUAL OF THE MOLLUSCA;* a Treatise on Recent and Fossil Shells. By Dr. S. P. WOODWARD, A.L.S. Fourth Edition. With Appendix by RALPH TATE, A.L.S., F.G.S. With numerous Plates and 300 Woodcuts. 6s. 6d. Cloth boards, 7s. 6d.
96. *ASTRONOMY.* By the late Rev. ROBERT MAIN, M.A. Third Edition, by WILLIAM THYNNE LYNN, B.A., F.R.A.S. 2s.
97. *STATICS AND DYNAMICS,* the Principles and Practice of; embracing also a clear development of Hydrostatics, Hydrodynamics, and Central Forces. By T. BAKER, C.E. 1s. 6d.
138. *TELEGRAPH,* Handbook of the; a Guide to.Candidates for Employment in the Telegraph Service. By R. BOND. Fourth Edition. Including Questions on Magnetism, Electricity, and Practical Telegraphy, by W. McGREGOR. 3s.‡
173. *PHYSICAL GEOLOGY,* partly based on Major-General PORT-LOCK'S "Rudiments of Geology." By RALPH TATE, A.L.S.,&c. Woodcuts. 2s.
174. *HISTORICAL GEOLOGY,* partly based on Major-General PORTLOCK'S "Rudiments." By RALPH TATE, A.L.S., &c. Woodcuts. 2s. 6d.
173 & 174. *RUDIMENTARY TREATISE ON GEOLOGY,* Physical and Historical. Partly based on Major-General PORTLOCK'S "Rudiments of Geology." By RALPH TATE, A.L.S., F.G.S., &c. In One Volume. 4s. 6d.‡
183 & 184. *ANIMAL PHYSICS,* Handbook of. By Dr. LARDNER, D.C.L., formerly Professor of Natural Philosophy and Astronomy in University College, Loud. With 520 Illustrations. In One Vol. 7s. 6d., cloth boards.
*** *Sold also in Two Parts, as follows :—*
183. ANIMAL PHYSICS. By Dr. LARDNER. Part I., Chapters I.—VII. 4s.
184. ANIMAL PHYSICS. By Dr. LARDNER. Part II., Chapters VIII.—XVIII. 3s.

FINE ARTS.

20. *PERSPECTIVE FOR BEGINNERS.* Adapted to Young Students and Amateurs in Architecture, Painting, &c. By GEORGE PYNE. 2s.
40. *GLASS STAINING, AND THE ART OF PAINTING ON GLASS.* From the German of Dr. GESSERT and EMANUEL OTTO FROMBERG. With an Appendix on THE ART OF ENAMELLING. 2s. 6d.
69. *MUSIC,* A Rudimentary and Practical Treatise on. With numerous Examples. By CHARLES CHILD SPENCER. 2s. 6d.
71. *PIANOFORTE,* The Art of Playing the. With numerous Exercises & Lessons from the Best Masters. By CHARLES CHILD SPENCER. 1s.6d.
69-71. *MUSIC & THE PIANOFORTE.* In one vol. Half bound, 5s.
181. *PAINTING POPULARLY EXPLAINED,* including Fresco, Oil, Mosaic, Water Colour, Water-Glass, Tempera, Encaustic, Miniature, Painting on Ivory, Vellum, Pottery, Enamel, Glass, &c. With Historical Sketches of the Progress of the Art by THOMAS JOHN GULLICK, assisted by JOHN TIMBS, F.S.A. Fifth Edition, revised and enlarged. 5s.‡
186. *A GRAMMAR OF COLOURING,* applied to Decorative Painting and the Arts. By GEORGE FIELD. New Edition, enlarged and adapted to the Use of the Ornamental Painter and Designer. By ELLIS A. DAVIDSON. With two new Coloured Diagrams, &c. 3s.‡
246. *A DICTIONARY OF PAINTERS, AND HANDBOOK FOR PICTURE AMATEURS;* including Methods of Painting, Cleaning, Relining and Restoring, Schools of Painting, &c. With Notes on the Copyists and Imitators of each Master. By PHILIPPE DARYL. 2s. 6d.‡

☞ *The ‡ indicates that these vols. may be had strongly bound at 6d. extra.*

7, STATIONERS' HALL COURT, LUDGATE HILL, E.C.

INDUSTRIAL AND USEFUL ARTS.

23. *BRICKS AND TILES*, Rudimentary Treatise on the Manufacture of. By E. DOBSON, M.R.I.B.A. Illustrated, 3s.‡
67. *CLOCKS, WATCHES, AND BELLS*, a Rudimentary Treatise on. By Sir EDMUND BECKETT, LL.D., Q.C. Seventh Edition, revised and enlarged. 4s. 6d. limp; 5s. 6d. cloth boards.
83**. *CONSTRUCTION OF DOOR LOCKS*. Compiled from the Papers of A. C. HOBBS, and Edited by CHARLES TOMLINSON, F.R.S. With Additions by ROBERT MALLET, M.I.C.E. Illus. 2s. 6d.
162. *THE BRASS FOUNDER'S MANUAL;* Instructions for Modelling, Pattern-Making, Moulding, Turning, Filing, Burnishing, Bronzing, &c. With copious Receipts, &c. By WALTER GRAHAM. 2s.‡
205. *THE ART OF LETTER PAINTING MADE EASY*. By J. G. BADENOCH. Illustrated with 12 full-page Engravings of Examples. 1s.
215. *THE GOLDSMITH'S HANDBOOK*, containing full Instructions for the Alloying and Working of Gold. By GEORGE E. GEE, 3s.‡
225. *THE SILVERSMITH'S HANDBOOK*, containing full Instructions for the Alloying and Working of Silver. By GEORGE E. GEE. 3s.‡
*** *The two preceding Works, in One handsome Vol., half-bound, entitled* "THE GOLDSMITH'S & SILVERSMITH'S COMPLETE HANDBOOK," 7s. [*Just published.*
224. *COACH BUILDING*, A Practical Treatise, Historical and Descriptive. By J. W. BURGESS. 2s. 6d.‡
235. *PRACTICAL ORGAN BUILDING*. By W. E. DICKSON, M.A., Precentor of Ely Cathedral. Illustrated. 2s. 6d.‡
249. *THE HALL-MARKING OF JEWELLERY PRACTICALLY CONSIDERED*. By GEORGE E. GEE. 3s.‡

MISCELLANEOUS VOLUMES.

36. *A DICTIONARY OF TERMS used in ARCHITECTURE, BUILDING, ENGINEERING, MINING, METALLURGY, ARCHÆOLOGY, the FINE ARTS, &c.* By JOHN WEALE. Fifth Edition. Revised by ROBERT HUNT, F.R.S. Illustrated. 5s. limp; 6s. cloth boards.
50. *THE LAW OF CONTRACTS FOR WORKS AND SERVICES*. By DAVID GIBBONS. Third Edition, enlarged. 3s.‡
112. *MANUAL OF DOMESTIC MEDICINE*. By R. GOODING, B.A., M.D. A Family Guide in all Cases of Accident and Emergency. 2s.‡
112*. *MANAGEMENT OF HEALTH*. A Manual of Home and Personal Hygiene. By the Rev. JAMES BAIRD, B.A. 1s.
150. *LOGIC*, Pure and Applied. By S. H. EMMENS. 1s. 6d.
153. *SELECTIONS FROM LOCKE'S ESSAYS ON THE HUMAN UNDERSTANDING*. With Notes by S. H. EMMENS. 2s.
154. *GENERAL HINTS TO EMIGRANTS*. Notices of the various Fields for Emigration, Hints on Outfits, Useful Recipes, &c. 2s.
157. *THE EMIGRANT'S GUIDE TO NATAL*. By ROBERT JAMES MANN, F.R.A.S., F.M.S. Second Edition. Map. 2s.
193. *HANDBOOK OF FIELD FORTIFICATION*, intended for the Guidance of Officers Preparing for Promotion. By Major W. W. KNOLLYS, F.R.G.S. With 163 Woodcuts. 3s.‡
194. *THE HOUSE MANAGER :* Being a Guide to Housekeeping. Practical Cookery, Pickling and Preserving, Household Work, Dairy Management, the Table and Dessert, Cellarage of Wines, Home-brewing and Wine-making, the Boudoir and Dressing-room, Travelling, Stable Economy, Gardening Operations, &c. By AN OLD HOUSEKEEPER. 3s. 6d.‡
194, *HOUSE BOOK (The)*. Comprising :—I. THE HOUSE MANAGER.
112 & By an OLD HOUSEKEEPER. II. DOMESTIC MEDICINE. By R. GOODING, M.D.
112*. III. MANAGEMENT OF HEALTH. By J. BAIRD. In One Vol., half-bound, 6s.

☞ *The ‡ indicates that these vols. may be had strongly bound at 6d. extra.*

LONDON : CROSBY LOCKWOOD AND SON,

EDUCATIONAL AND CLASSICAL SERIES.

HISTORY.

1. **England, Outlines of the History of;** more especially with reference to the Origin and Progress of the English Constitution. By WILLIAM DOUGLAS HAMILTON, F.S.A., of Her Majesty's Public Record Office. 4th Edition, revised. 5s.; cloth boards, 6s.
5. **Greece, Outlines of the History of;** in connection with the Rise of the Arts and Civilization in Europe. By W. DOUGLAS HAMILTON, of University College, London, and EDWARD LEVIEN, M.A., of Balliol College, Oxford. 2s. 6d.; cloth boards, 3s. 6d.
7. **Rome, Outlines of the History of:** from the Earliest Period to the Christian Era and the Commencement of the Decline of the Empire. By EDWARD LEVIEN, of Balliol College, Oxford. Map, 2s. 6d.; cl. bds. 3s. 6d.
9. **Chronology of History, Art, Literature, and Progress,** from the Creation of the World to the Present Time. The Continuation by W. D. HAMILTON, F.S.A. 3s.; cloth boards, 3s. 6d.
50. **Dates and Events in English History,** for the use of Candidates in Public and Private Examinations. By the Rev. E. RAND. 1s.

ENGLISH LANGUAGE AND MISCELLANEOUS.

11. **Grammar of the English Tongue, Spoken and Written.** With an Introduction to the Study of Comparative Philology. By HYDE CLARKE, D.C.L. Fourth Edition. 1s. 6d.
11*. **Philology:** Handbook of the Comparative Philology of English, Anglo-Saxon, Frisian, Flemish or Dutch, Low or Platt Dutch, High Dutch or German, Danish, Swedish, Icelandic, Latin, Italian, French, Spanish, and Portuguese Tongues. By HYDE CLARKE, D.C.L. 1s.
12. **Dictionary of the English Language,** as Spoken and Written. Containing above 100,000 Words. By HYDE CLARKE, D.C.L. 3s. 6d.; cloth boards, 4s. 6d.; complete with the GRAMMAR, cloth bds., 5s. 6d.
48. **Composition and Punctuation,** familiarly Explained for those who have neglected the Study of Grammar. By JUSTIN BRENAN. 17th Edition. 1s. 6d.
49. **Derivative Spelling-Book:** Giving the Origin of Every Word from the Greek, Latin, Saxon, German, Teutonic, Dutch, French, Spanish, and other Languages; with their present Acceptation and Pronunciation. By J. ROWBOTHAM, F.R.A.S. Improved Edition. 1s. 6d.
51. **The Art of Extempore Speaking:** Hints for the Pulpit, the Senate, and the Bar. By M. BAUTAIN, Vicar-General and Professor at the Sorbonne. Translated from the French. 8th Edition, carefully corrected. 2s. 6d.
52. **Mining and Quarrying,** with the Sciences connected therewith. First Book of, for Schools. By J. H. COLLINS, F.G.S., Lecturer to the Miners' Association of Cornwall and Devon. 1s.
53. **Places and Facts in Political and Physical Geography,** for Candidates in Examinations. By the Rev. EDGAR RAND, B.A. 1s.
54. **Analytical Chemistry,** Qualitative and Quantitative, a Course of. To which is prefixed, a Brief Treatise upon Modern Chemical Nomenclature and Notation. By WM. W. PINK and GEORGE E. WEBSTER. 2s.

THE SCHOOL MANAGERS' SERIES OF READING BOOKS,

Edited by the Rev. A. R. GRANT, Rector of Hitcham, and Honorary Canon of Ely; formerly H.M. Inspector of Schools.

INTRODUCTORY PRIMER, 3d.

	s. d.		s. d.
FIRST STANDARD	0 6	FOURTH STANDARD	1 2
SECOND ,,	0 10	FIFTH ,,	1 6
THIRD ,,	1 0	SIXTH ,,	1 6

LESSONS FROM THE BIBLE. Part I. Old Testament. 1s.
LESSONS FROM THE BIBLE. Part II. New Testament, to which is added THE GEOGRAPHY OF THE BIBLE, for very young Children. By Rev. C THORNTON FORSTER. 1s. 2d. *** Or the Two Parts in One Volume. 2s

7, STATIONERS' HALL COURT, LUDGATE HILL, E.C.

FRENCH.

24. **French Grammar.** With Complete and Concise Rules on the Genders of French Nouns. By G. L. STRAUSS, Ph.D. 1s. 6d.
25. **French-English Dictionary.** Comprising a large number of New Terms used in Engineering, Mining, &c. By ALFRED ELWES. 1s. 6d.
26. **English-French Dictionary.** By ALFRED ELWES. 2s.
25,26. **French Dictionary** (as above). Complete, in One Vol., 3s.; cloth boards, 3s. 6d. *** Or with the GRAMMAR, cloth boards, 4s. 6d.
47. **French and English Phrase Book:** containing Introductory Lessons, with Translations, several Vocabularies of Words, a Collection of suitable Phrases, and Easy Familiar Dialogues. 1s. 6d.

GERMAN.

39. **German Grammar.** Adapted for English Students, from Heyse's Theoretical and Practical Grammar, by Dr. G. L. STRAUSS. 1s. 6d.
40. **German Reader:** A Series of Extracts, carefully culled from the most approved Authors of Germany; with Notes, Philological and Explanatory. By G. L. STRAUSS, Ph.D. 1s.
41-43. **German Triglot Dictionary.** By N. E. S. A. HAMILTON. In Three Parts. Part I. German-French-English. Part II. English-German-French. Part III. French-German English. 3s., or cloth boards, 4s.
41-43 & 39. **German Triglot Dictionary** (as above), together with German Grammar (No. 39), in One Volume, cloth boards, 5s.

ITALIAN.

27. **Italian Grammar,** arranged in Twenty Lessons, with a Course of Exercises. By ALFRED ELWES. 1s. 6d.
28. **Italian Triglot Dictionary,** wherein the Genders of all the Italian and French Nouns are carefully noted down. By ALFRED ELWES. Vol. 1. Italian-English-French. 2s. 6d.
30. **Italian Triglot Dictionary.** By A. ELWES. Vol. 2. English-French-Italian. 2s. 6d.
32. **Italian Triglot Dictionary.** By ALFRED ELWES. Vol. 3. French-Italian-English. 2s. 6d.
28,30,32. **Italian Triglot Dictionary** (as above). In One Vol., 7s. 6d. Cloth boards.

SPANISH AND PORTUGUESE.

34. **Spanish Grammar,** in a Simple and Practical Form. With a Course of Exercises. By ALFRED ELWES. 1s. 6d.
35. **Spanish-English and English-Spanish Dictionary** Including a large number of Technical Terms used in Mining, Engineering, &c. with the proper Accents and the Gender of every Noun. By ALFRED ELWES 4s.; cloth boards, 5s. *** Or with the GRAMMAR, cloth boards, 6s.
55. **Portuguese Grammar,** in a Simple and Practical Form. With a Course of Exercises. By ALFRED ELWES. 1s. 6d.
56. **Portuguese-English and English-Portuguese Dictionary.** Including a large number of Technical Terms used in Mining, Engineering, &c., with the proper Accents and the Gender of every Noun. By ALFRED ELWES. 5s.; cloth boards, 6s. *** Or with the GRAMMAR, cloth boards, 7s.

HEBREW.

46*. **Hebrew Grammar.** By Dr. BRESSLAU. 1s. 6d.
44. **Hebrew and English Dictionary,** Biblical and Rabbinical; containing the Hebrew and Chaldee Roots of the Old Testament Post-Rabbinical Writings. By Dr. BRESSLAU. 6s.
46. **English and Hebrew Dictionary.** By Dr. BRESSLAU. 3s.
44,46. **Hebrew Dictionary** (as above), in Two Vols., complete, with
46*. the GRAMMAR, cloth boards. 12s.

LONDON: CROSBY LOCKWOOD AND SON,

LATIN.

19. **Latin Grammar.** Containing the Inflections and Elementary Principles of Translation and Construction. By the Rev. THOMAS GOODWIN, M.A., Head Master of the Greenwich Proprietary School. 1s. 6d.
20. **Latin-English Dictionary.** By the Rev. THOMAS GOODWIN, M.A. 2s.
22. **English-Latin Dictionary;** together with an Appendix of French and Italian Words which have their origin from the Latin. By the Rev. THOMAS GOODWIN, M.A. 1s. 6d.
20,22. **Latin Dictionary** (as above). Complete in One Vol., 3s. 6d. cloth boards, 4s. 6d. *.* Or with the GRAMMAR, cloth boards, 5s. 6d.

LATIN CLASSICS. With Explanatory Notes in English.
1. **Latin Delectus.** Containing Extracts from Classical Authors, with Genealogical Vocabularies and Explanatory Notes, by H. YOUNG. 1s. 6d.
2. **Cæsaris Commentarii de Bello Gallico.** Notes, and a Geographical Register for the Use of Schools, by H. YOUNG. 2s.
3. **Cornelius Nepos.** With Notes. By H. YOUNG. 1s.
4. **Virgilii Maronis Bucolica et Georgica.** With Notes on the Bucolics by W. RUSHTON, M.A., and on the Georgics by H. YOUNG. 1s. 6d.
5. **Virgilii Maronis Æneis.** With Notes, Critical and Explanatory, by H. YOUNG. New Edition, revised and improved With copious Additional Notes by Rev. T. H. L. LEARY, D.C.L., formerly Scholar of Brasenose College, Oxford. 3s.
5*. ——— Part 1. Books i.—vi., 1s. 6d.
5**. ——— Part 2. Books vii.—xii., 2s.
6. **Horace;** Odes, Epode, and Carmen Sæculare. Notes by H. YOUNG. 1s. 6d.
7. **Horace;** Satires, Epistles, and Ars Poetica. Notes by W. BROWNRIGG SMITH, M.A., F.R.G.S. 1s. 6d.
8. **Sallustii** Crispi Catalina et Bellum Jugurthinum. Notes, Critical and Explanatory, by W. M. DONNE, B.A., Trin. Coll., Cam. 1s. 6d.
9. **Terentii Andria et Heautontimorumenos.** With Notes, Critical and Explanatory, by the Rev. JAMES DAVIES, M.A. 1s. 6d.
10. **Terentii Adelphi, Hecyra, Phormio.** Edited, with Notes, Critical and Explanatory, by the Rev. JAMES DAVIES, M.A. 2s.
11. **Terentii Eunuchus, Comœdia.** Notes, by Rev. J. DAVIES, M.A. 1s. 6d.
12. **Ciceronis Oratio pro Sexto Roscio Amerino.** Edited, with an Introduction, Analysis, and Notes, Explanatory and Critical, by the Rev. JAMES DAVIES, M.A. 1s. 6d.
13. **Ciceronis Orationes in Catilinam, Verrem, et pro Archia.** With Introduction, Analysis, and Notes, Explanatory and Critical, by Rev. T. H. L. LEARY, D.C.L. formerly Scholar of Brasenose College, Oxford. 1s. 6d.
14. **Ciceronis Cato Major, Lælius, Brutus, sive de Senectute, de Amicitia, de Claris Oratoribus Dialogi.** With Notes by W. BROWNRIGG SMITH, M.A., F.R.G.S.
16. **Livy:** History of Rome. Notes by H. YOUNG and W. B. SMITH, M.A. Part 1. Books i., ii., 1s. 6d.
16*. ——— Part 2. Books iii., iv., v., 1s. 6d.
17. ——— Part 3. Books xxi., xxii., 1s. 6d.
19. **Latin Verse Selections,** from Catullus, Tibullus, Propertius, and Ovid. Notes by W. B. DONNE, M.A., Trinity College, Cambridge. 2s.
20. **Latin Prose Selections,** from Varro, Columella, Vitruvius, Seneca, Quintilian, Florus, Velleius Paterculus, Valerius Maximus Suetonius, Apuleius, &c. Notes by W. B. DONNE, M.A. 2s.
21. **Juvenalis Satiræ.** With Prolegomena and Notes by T. H. S. ESCOTT, B.A., Lecturer on Logic at King's College, London. 2s.

7, STATIONERS' HALL COURT, LUDGATE HILL, E.C.

GREEK.

14. **Greek Grammar**, in accordance with the Principles and Philological Researches of the most eminent Scholars of our own day. By HANS CLAUDE HAMILTON. 1s. 6d.

15,17. **Greek Lexicon.** Containing all the Words in General Use, with their Significations, Inflections, and Doubtful Quantities. By HENRY R. HAMILTON. Vol. 1. Greek-English, 2s. 6d.; Vol. 2. English-Greek, 2s. Or the Two Vols. in One, 4s. 6d.: cloth boards, 5s.

14,15. **Greek Lexicon** (as above). Complete, with the GRAMMAR, in 17. One Vol., cloth boards, 6s.

GREEK CLASSICS. With Explanatory Notes in English.

1. **Greek Delectus.** Containing Extracts from Classical Authors, with Genealogical Vocabularies and Explanatory Notes, by H. YOUNG. New Edition, with an improved and enlarged Supplementary Vocabulary, by JOHN HUTCHISON, M.A., of the High School, Glasgow. 1s. 6d.

2, 3. **Xenophon's Anabasis;** or, The Retreat of the Ten Thousand. Notes and a Geographical Register, by H. YOUNG. Part 1. Books i. to iii., 1s. Part 2. Books iv. to vii., 1s.

4. **Lucian's Select Dialogues.** The Text carefully revised, with Grammatical and Explanatory Notes, by H. YOUNG. 1s. 6d.

5-12. **Homer**, The Works of. According to the Text of BAEUMLEIN. With Notes, Critical and Explanatory, drawn from the best and latest Authorities, with Preliminary Observations and Appendices, by T. H. L. LEARY, M.A., D.C.L.

THE ILIAD: Part 1. Books i. to vi., 1s. 6d. | Part 3. Books xiii. to xviii., 1s. 6d.
Part 2. Books vii. to xii., 1s. 6d. | Part 4. Books xix. to xxiv., 1s. 6d.
THE ODYSSEY: Part 1. Books i. to vi., 1s. 6d | Part 3. Books xiii. to xviii., 1s. 6d.
Part 2. Books vii. to xii., 1s. 6d. | Part 4. Books xix. to xxiv., and Hymns, 2s.

13. **Plato's Dialogues:** The Apology of Socrates, the Crito, and the Phædo. From the Text of C. F. HERMANN. Edited with Notes, Critical and Explanatory, by the Rev. JAMES DAVIES, M.A. 2s.

14-17. **Herodotus**, The History of, chiefly after the Text of GAISFORD. With Preliminary Observations and Appendices, and Notes, Critical and Explanatory, by T. H. L. LEARY, M.A., D.C.L.
Part 1. Books i., ii. (The Clio and Euterpe), 2s.
Part 2. Books iii., iv. (The Thalia and Melpomene), 2s.
Part 3. Books v.-vii. (The Terpsichore, Erato, and Polymnia), 2s.
Part 4. Books viii., ix. (The Urania and Calliope) and Index, 1s. 6d.

18. **Sophocles:** Œdipus Tyrannus. Notes by H. YOUNG. 1s.
20. **Sophocles:** Antigone. From the Text of DINDORF. Notes, Critical and Explanatory, by the Rev. JOHN MILNER, B.A. 2s.
23. **Euripides:** Hecuba and Medea. Chiefly from the Text of DINDORF. With Notes, Critical and Explanatory, by W. BROWNRIGG SMITH, M.A., F.R.G.S. 1s. 6d.
26. **Euripides:** Alcestis. Chiefly from the Text of DINDORF. With Notes, Critical and Explanatory, by JOHN MILNER, B.A. 1s. 6d.
30. **Æschylus:** Prometheus Vinctus: The Prometheus Bound. From the Text of DINDORF. Edited, with English Notes, Critical and Explanatory, by the Rev. JAMES DAVIES, M.A. 1s.
32. **Æschylus:** Septem Contra Thebes: The Seven against Thebes. From the Text of DINDORF. Edited, with English Notes, Critical and Explanatory, by the Rev. JAMES DAVIES, M.A. 1s.
40. **Aristophanes:** Acharnians. Chiefly from the Text of C. H. WEISE. With Notes, by C. S. T. TOWNSHEND, M.A. 1s. 6d.
41. **Thucydides:** History of the Peloponnesian War. Notes by H. YOUNG. Book 1. 1s. 6d.
42. **Xenophon's Panegyric on Agesilaus.** Notes and Introduction by LL. F. W. JEWITT. 1s. 6d.
43. **Demosthenes.** The Oration on the Crown and the Philippics. With English Notes. By Rev. T. H. L. LEARY, D.C.L., formerly Scholar of Brasenose College, Oxford. 1s. 6d.

CROSBY LOCKWOOD AND SON, 7, STATIONERS' HALL COURT, E.C.

www.ingramcontent.com/pod-product-compliance
Lightning Source LLC
Chambersburg PA
CBHW020825230426
43666CB00007B/1103